PROBABILITY ADVENTURES

VERSION 1.0

BY MARK HUBER

©2021 Mark Huber
Version 1.0

Graphics

Front Cover

   Image by Stefan Keller from Pixabay.

Chapter 1

   Image by Yale Art Gallery (https://artgallery.yale.edu/collections/objects/170864)

   Image by Yale Art Gallery (https://artgallery.yale.edu/collections/objects/6521)

Chapter 5

   Image by Clker-Free-Vector-Images from Pixabay.

Chapter 6

   Image by Gordon Johnson from Pixabay.

Chapter 11

   Tavern image by Clker-Free-Vector-Images from Pixabay.

   Fish image by Syaibatul Hamdi from Pixabay.

Chapter 12

   Image by Adnan Khalid from Pixabay.

Chapter 13

   Image by Thanasis Papazacharias from Pixabay.

Chapter 14

   Image by Mark Huber.

Chapter 15

   Image by Markus Steidle from Pixabay.

Chapter 16

   Image by Mark Huber.

Chapter 17

   Image by Gerd Altmann from Pixabay.

Chapter 18

   Image by kevinp133 from Pixabay

Chapter 19

   Image by Shutterbug75 from Pixabay.

Chapter 20

   Image by Devanath from Pixabay.

Chapter 21

   Image by Rebecca Matthews from Pixabay.

Chapter 22

   Image by Mark Huber.

Chapter 23

   Image by Jalyn Bryce from Pixabay.

Chapter 24

   Image by Borkia from Pixabay.

Chapter 25

   Image by Paul Barlow from Pixabay.

Chapter 26

   Image by Mark Huber.

Chapter 28

   Image by Mark Huber.

Contents

1 What is Probability?   1
Probability & Logic   1
    True and False   1
    Indicator functions   1
    Probability functions   2
Assigning Probabilities   3
Mathematical probability   3
    Sets   3
    Coins and Dice   4
Encounters   4
    1. Basic indicators   4
    2. More basic indicators . .   4
    3. Basic probabilities . . .   4
    4. More basic probabilities   4
    5. Indicator functions . . .   5
    6. Indicator function factors   5
    7. Graphing indicator functions   5
    8. Graphing indication function factors . . .   5
    9. Disjoint reals   5
    10. Disjoint integers   5
    11. Truth or dare?   5
    12. Truth or dare, arithmetic edition   5
    13. Counting true statements   5
    14. Adding indicators . . .   5
    15. Two dice   5
    16. Three dice   5
    17. Principle of Indifference   5
    18. Principle of Indifference for three dice   5

2 Logical Operators   6
Modifying logical statements . .   6
    Logical OR   6
    Logical AND   7
    Logical NOT   7
Order of Operations   7
    Logic and arithmetic   8
De Morgan's Laws   9
Solving the story 10
Implication 10
    Subsets 10
Encounters 10
    19. Logical AND, OR, and NOT 10
    20. Negation 10
    21. Countable logical AND 11
    22. Countable logical OR . 11
    23. Logic to arithmetic . . 11
    24. More logic to arithmetic 11
    25. Proof with indicator functions 11
    26. More proof with indicator functions 11
    27. Logical order of operations 11
    28. More logical order of operations 11
    29. De Morgan's Laws . . . 11
    30. More De Morgan's Laws 11
    31. Implication and subsets 11
    32. More implications and subsets 11

3 The Rules of Probability   12
Mathematical probability 12
    The collection of events . . 12
    The probability function . . 13
Five more rules of probability . 13
    Probability of a false statement is 0 14
    Finite disjoint events 14
Negation rule 15
Inclusion-Exclusion 15
    Implication 16
Solving the story 16
Examples 16
Encounters 17
    33. True statements 17
    34. More true statements . 17
    35. False statements . . . 17
    36. More false statements 17
    37. Rolling the dice 17
    38. A fair six-sided die . . . 17
    39. Using rules 17

40. Breaking apart problems	18
41. Negation rule	18
42. More negation	18
43. Implication	18
44. More implication	18
45. Combining rules	18
46. Inclusion-Exclusion	18

4 Conditional Probability — 19
Partial Information 19
Solving The Problem of Points . 20
The general formula 20
    Two-stage experiments . . 21
Odds 22
Encounters 22
    47. A conditional die 22
    48. Basic conditional probability 22
    49. Interval conditioning . 22
    50. Another conditional die 22
    51. Blood tests 22
    52. Looking for copper . . . 22
    53. The satellite 23
    54. Emergency! 23
    55. Raining once more . . 23
    56. Tumor odds 23
    57. Odds and fairness . . . 23
    58. Odds and probabilities 23
    59. Experience 23
    60. Drawing cards 23
    61. The food pantry 23
    62. A corny problem 23

5 Independence — 24
Independence of two events . . 24
More than two events 25
    Sequences 25
Solving the story 26
    Note on independence of three or more events 26
Encounters 27
    63. Independence of two events 27
    64. Independence of three events 27
    65. All false 27
    66. Gold! 27
    67. The archers 27
    68. Finding a vaccine . . . 27
    69. Factory woes 27
    70. High School success . 27

6 Bayes' Rule — 28
Flipping conditioning 28
    Solving the story 29
Using Bayes' Rule 29
    Terminology 30
The Law of Total Probability . . 31
Encounters 31
    71. Bayes' Rule 31
    72. Probability vectors . . . 32
    73. Machine problems . . . 32
    74. Building cars 32
    75. Cholesterol 32
    76. Genetics 32
    77. A bit of a screw-up . . 32
    78. Getting in a bind . . . 32

7 Random variables — 33
Random variables 33
    Solving the story 34
Independence of random variables 35
σ-algebras of sets 36
Continuous random variables . 37
Encounters 37
    79. Ten independent events 37
    80. Different rolls 37
    81. Flips of a 0-1 coin . . . 37
    82. Conditioning on sums 37
    83. Six arrows 38
    84. Ozone 38
    85. Coin flips 38
    86. Scouting for water . . . 38
    87. Measurable sets 38
    88. More measurable sets . 38
    89. The triangle 38
    90. The rectangle 38
    91. Transport 38
    92. Manufacturing time . . 38

8 Uniform random variables — 39
The Discrete Uniform distribution 39
    Intersection of sets 39
The Continuous Uniform distribution 40
    The standard uniform . . . 41
Uniform properties 41

  Solving the Story 42
 Encounters 42
  93. Summing three dice . . 42
  94. More sums of three dice 42
  95. Standard uniform . . . 42
  96. Wider uniform 42
  97. Functions of a uniform 42
  98. Absolutely uniform . . 42
  99. Conditional uniform . 42
  100. More conditional
    uniforms 42
  101. Defective testing . . . 42
  102. A deeper test 42

9 Functions of random variables 43
 Functions of random variables 43
  The cumulative distribution
   function 43
  Functions keep or destroy
   information 44
 The Bernoulli distribution . . . 44
 The exponential distribution . . 45
  Solving the story problem . 45
 The Fundamental Theorem of
  Simulation 46
 The self-replicating uniform . . 46
 Encounters 46
  103. Finding a cdf 46
  104. Finding another cdf . 46
  105. Indicators of a uniform 46
  106. Graphing the cdf . . . 46
  107. Scaling and shifting
    uniforms 47
  108. Exponential probabilities 47
  109. Scaling exponentials . 47
  110. A triple problem . . . 47
  111. Conditional expectations 47
  112. What does iid mean? . 47
  113. The ceiling function . 47
  114. Continuous to discrete 47

10 The Bernoulli Process 48
 Stochastic processes 48
 Number of successes on n trials 49
  Solving the Story 49
 Number of trials until the 1st
  success 49

 Number of trials until the rth
  success 50
 The picture 50
 Encounters 51
  115. Basic binomials 51
  116. Tails of binomials . . . 51
  117. Basic geometrics . . . 51
  118. More geometrics . . . 51
  119. Negative binomials . . 51
  120. More negative binomials 51
  121. The drug trial 51
  122. Waiting for success . 51
  123. The bad batch 51
  124. Conditioned on failure 51
  125. Fly away 51
  126. Looking for medicine . 51

11 Poisson point processes 53
 Creating a BPP using geometrics 53
 Exponential random variables . 53
 Using Exponentials to create a
  Poisson point process . . . 54
 Three questions 54
  Exponential 55
  Gamma 55
  Poisson 55
 Encounters 56
  127. Tails of exponentials . 56
  128. Conditional exponentials 56
  129. Bernoullis from
    geometrics 56
  130. More Bernoullis from
    geometrics 56
  131. Understanding Poisson
    point processes . . . 56
  132. Features of PPP . . . 56
  133. Conditional PPP . . . 57
  134. More conditional PPP 57
  135. The restaurant 57
  136. On the factory floor . 57

12 Densities for continuous random variables 58
 Differentials 58
 Densities 59
 How the cdf and pdf relate . . . 60
  Moving from the pdf to the
   cdf 60

  Moving from the cdf to the pdf 61
Encounters 62
  137. Density of a uniform . 62
  138. Gamma density . . . 62
  139. More uniform density 62
  140. Exponential density . 62
  141. Another exponential density 62
  142. Cdf of a gamma . . . 62
  143. Normalizing an exponential 62
  144. Normalizing a beta density 62
  145. Density of a function of a uniform 62
  146. Shifting and scaling a uniform 62
  147. From cdf to pdf 62
  148. Double exponential . 62

13 Densities for discrete random variables   63
Densities against counting measure 63
Discrete cdf functions 64
Measures of central tendency . 64
  The mode 64
  The median 65
Shifting and scaling 66
The survival function 67
Encounters 67
  149. Pdf for a die 67
  150. Cdf for a die 67
  151. Binomial mode 67
  152. Gamma mode 67
  153. Gamma median . . . 67
  154. Median sets of a die . 67
  155. Cdf of a scaled die . . 67
  156. Pdf of a scaled die . . 67
  157. Scaling a beta 67
  158. Scaled and shifted gamma 67
  159. Survival function of an exponential . . . 67
  160. Survival function of a die 67

14 Mean of a random variable   68
The sample average 68
  Moving to the infinite . . . 69
The Strong Law of Large Numbers 69
  Events of probability 1 . . 70
Linearity of expectation 70
Symmetry 71
Writing sums as integrals . . . 71
Encounters 72
  161. Mean of finite random variables 72
  162. Another mean of a finite random variable 72
  163. Mean of a discrete uniform 72
  164. Another mean of a discrete uniform . . 72
  165. Mean of a die roll . . 72
  166. Mean of a sum 72
  167. Mean of a difference . 72
  168. Mean of a function of a random variable . 72
  169. Symmetry of a discrete uniform 72
  170. A symmetric finite random variable . . 72
  171. Vertigon's Army . . . 72
  172. A paper mill 72
  173. Four stores 72
  174. Streamlining 73
  175. The SLLN in action . 73
  176. To infinity! 73
  177. Two tasks 73
  178. A weighty matter . . . 73

15 Mean of a general random variable   74
Expectations using densities . . 74
  Solving the story 76
Symmetry 76
Monte Carlo Methods 76
Encounters 76
  179. Mean of an exponential 76
  180. Mean of a uniform . . 77
  181. Mean of functions of a continuous uniform 77
  182. Mean of a general exponential 77

183. Monte Carlo with uniforms 77
184. Monte Carlo with exponentials 77
185. Polynomials of continuous uniforms 77
186. Polynomials of discrete uniforms 77
187. A nonintegrable density 77
188. A nonintegrable random variable . . 77
189. Deriving formulas . . 77
190. A cubic formula . . . 77

16 Conditional expectation 78
Knowing about a random variable 78
The Fundamental Theorem of Probability 79
Solving the Story 79
Properties of conditional expectation 79
Expectation and Probability Trees 80
The expectation of a Geometric random variable 81
Connection to conditional probability 81
Encounters 82
191. Raising sales 82
192. Fire! 82
193. Food for thought . . . 82
194. The sell out 82
195. Random numbers of dice 82
196. Random number of exponentials 83
197. The Bayesian, Part I . 83
198. The Bayesian, Part II 83
199. Using the FTP 83
200. More FTP 83

17 Joint Densities of Random Variables 84
Univariate random variables . . 84
Densities in higher dimensions 84
Solving the story 85
Marginal densities 86
Showing independence using densities 87
Encounters 88
201. Joint densities 88
202. More joint densites . 88
203. Independence 88
204. More independence . 88
205. Multiple integrals . . 88
206. Running the triangle 88
207. Discrete joint densities 88
208. Joint discrete uniforms 89
209. Factoring continuous densities 89
210. Factoring discrete densities 89
211. Dependent joint continuous uniforms 89
212. Joint continuous uniforms 89

18 Random variables as vector spaces 90
Vector space 90
Inner products 91
Norms 91
Calculating the covariance and standard deviation 92
Solving the story 92
Proof covariance is an inner product 93
Encounters 94
213. Rules of inner products 94
214. More rules of inner products 94
215. Rules of inner product norms 94
216. Standard deviation . 94
217. Standard deviation of discrete random variables 94
218. Discrete via density . 94
219. Covariance of a joint uniform 94
220. A uniform triangle . . 94
221. Continuous uniform part I 94
222. Continuous uniform part II 94

19 Correlation — 96
- Correlation 96
  - Solving the Story 96
- Bounds on the correlation . . . 97
- Uncorrelated 97
- Properties of variance and covariance. 98
- Proof of the Cauchy Schwartz inequality. 99
- Encounters 99
  - 223. Affine transforms . . 99
  - 224. A bit of negativity . . 99
  - 225. Correlation of discrete uniforms 100
  - 226. Correlation of independent random variables . . 100
  - 227. Understanding functions 100
  - 228. Another negative result 100
  - 229. The Pythagorean Theorem 100
  - 230. More Pythagoras . . . 100
  - 231. Sample averages . . . 100
  - 232. How many to average? 100

20 Adding random variables — 101
- Adding discrete random variables 101
- Independent Random variables 102
  - Solving the story 102
- Using polynomials to add discrete random variables . 104
- Encounters 105
  - 233. Adding random variables 105
  - 234. More addition 105
  - 235. Adding independent dice 105
  - 236. More dice 105
  - 237. Adding continuous random variables . . 105
  - 238. More continuous addition 105
  - 239. Generating function of a Poisson 105
  - 240. Adding Poissons . . . 106
  - 241. Wolfram Alpha to the rescue 106
  - 242. Adding lots of dice . . 106

21 The Central Limit Theorem — 107
- The moment generating function 107
- Shifting and scaling 108
- The mgf and the sum of random variables 108
  - The normal distribution . . 108
  - Solving the story 109
  - The half integer correction 109
- Encounters 110
  - 243. CLT for exponentials 110
  - 244. CLT for uniforms . . . 110
  - 245. CLT for geometrics . . 110
  - 246. CLT for dice 110
  - 247. Another exponential CLT 110
  - 248. CLT for unknown distributions . . . 110
  - 249. Factory accidents . . 110
  - 250. Customer arrivals . . 110
  - 251. Checking the CLT . . 110
  - 252. Prescriptions 110

22 Normal random variables — 111
- Shifting and scaling standard normals 111
  - Solving the story 111
- Adding independent normals . 112
- Multivariate normals 112
  - Covariance for the multivariate normal 112
- Density of a multivariate normal 113
- Encounters 113
  - 253. Shifting and scaling normals 113
  - 254. More shifting and scaling normals . . . 113
  - 255. Adding independent normals 113
  - 256. Adding three independent normals 114
  - 257. Multivariate scaling and shifting 114
  - 258. Finding multivariate normal parameters . 114

23 Bayes' rule for densities — 115
- Differentials and Bayes 115
  - Solving the story 116

The beta distribution 116
Encounters 117
    259. Archytas Medical . . . 117
    260. Updating rate of an exponential 117
    261. Toothpaste troubles . 117
    262. Large and small orders 117
    263. Ronco Survey Group 117
    264. Some rocket science . 117
    265. Dimer Medicine . . . 118
    266. Torrent Music 118
    267. The assembly line . . 118
    268. Inspector Quinn . . . 118
    269. Rush Hour 118
    270. Uniform Bayes 118

24 The multinomial distribution 119
More than two outcomes 119
Density of the multinomial distribution 120
Moments of multinomial random variables 122
    Covariance of multinomial 122
    Solving the Alchemist's story 123
Encounters 123
    271. Probabilities for multinomials 123
    272. More probabilities of multinomials 123
    273. Means of multinomials 123
    274. Multinomials means . 123
    275. Multinomial covariance 123
    276. More covariances of multinomials 123

25 Tail inequalities 124
Expectation as balance point . 124
    Solving the Story 125
Chebyshev's inequality 125
Sample Averages 126
    Relative Error 126
Encounters 127
    277. Markov's inequality . 127
    278. More Markov's inequality 127
    279. Chebyshev's inequality 127
    280. More Chebyshev's inequality 127
    281. Chebyshev via standard deviation 127
    282. More Chebyshev via standard deviation . 127
    283. Sample averages and Chebyshev 128
    284. More sample averages and Chebyshev . . . 128
    285. Construction woes . . 128
    286. The vaccine 128

26 Chernoff inequalities 129
The Chernoff inequality 129
    Solving the story 130
Sums of random variables . . . 130
    Sample averages 131
Encounters 131
    287. Chernoff for a Poisson 131
    288. More Chernoff for Poissons 131
    289. Chernoff for gammas 131
    290. More Chernoff for gammas 131
    291. Chernoff given the mgf 131
    292. More Chernoff given the mgf 131
    293. Chernoff for uniforms 132
    294. Chernoff for discrete uniforms 132

27 The hypergeometric distribution 133
The Hypergeometric distribution 133
    Solving the Story 134
Uniform points 134
Moments of hypergeometrics . . 135
Another view 136
Encounters 137
    295. Those crazy eights . . 137
    296. Pick a card, any card 137
    297. A bag of marbles . . . 137
    298. Strains 137
    299. A box of screws 137
    300. A bag of tokens 137
    301. The mission 137
    302. Back to the assembly line 138
    303. The forest 138

304. Turtles all the way down 138

28 Poisson point processes over general spaces   139
The uniform point of view . . . 139
    From uniform to binomial . 140
Encounters 140
    305. Back to the cellar . . 140
    306. The disease 140
    307. The restaurant 140
    308. The book 140
    309. An abstract space . . 141
    310. More abstract fun! . . 141

29 Transforming multivariate random variables   142
Multivariate transformations . . 142
    Nonlinear transformations in one dimension . . 142
    Higher dimensional transforms 143
Solving the story 144
Polar coordinates 144
Encounters 145
    311. One dimensional transform 145
    312. Transforming an exponential 145
    313. Two dimensional transform 145
    314. Another 2D transform 145

30 Encounters resolved   146
What is Probability? 146
    1. Basic indicators 146
    3. Basic probabilities . . . 146
    5. Indicator functions . . 146
    7. Graphing indicator functions 146
    9. Disjoint reals 146
    11. Truth or dare? 147
    13. Counting true statements 147
    15. Two dice 147
    17. Principle of Indifference 147
Logical Operators 147
    19. Logical AND, OR, and NOT 147
    21. Countable logical AND 147
    23. Logic to arithmetic . . 147
    25. Proof with indicator functions 148
    27. Logical order of operations 148
    29. De Morgan's Laws . . . 148
    31. Implication and subsets 148
Rules of Probability 148
    33. True statements 148
    35. False statements . . . 149
    37. Rolling the dice 149
    39. Using rules 149
    41. Negation rule 149
    43. Implication 149
    45. Combining rules 149
Conditional Probability 149
    47. A conditional die 149
    49. Interval conditioning . 150
    51. Blood tests 150
    53. The satellite 150
    55. Raining once more . . 150
    57. Odds and fairness . . . 150
    59. Experience 151
    61. The food pantry . . . 151
Independence 151
    63. Independence of two events 151
    65. All false 151
    67. The archers 152
    69. Factory woes 152
Bayes Rule 152
    71. Bayes' Rule 152
    73. Machine problems . . . 152
    75. Cholesterol 153
    77. A bit of a screw-up . . 153
Random Variables 154
    79. Ten independent events 154
    81. Flips of a 0-1 coin . . . 154
    83. Six arrows 154
    85. Coin flips 155
    87. Measurable sets 155
    89. The triangle 155
    91. Transport 155
Uniform random Variables . . . 156
    93. Summing three dice . . 156
    95. Standard uniform . . . 156
    97. Functions of a uniform 156
    99. Conditional uniform . 156
    101. Defective testing . . . 157
Functions of random Variables   157

- 103. Finding a cdf 157
- 105. Indicators of a uniform 157
- 107. Scaling and shifting uniforms 157
- 109. Scaling exponentials . 158
- 111. Conditional expectations 158
- 113. The ceiling function . 158

The Bernoulli Process 159
- 115. Basic binomials . . . 159
- 117. Basic geometrics . . . 159
- 119. Negative binomials . . 159
- 121. The drug trial 159
- 123. The bad batch . . . 159
- 125. Fly away 160

The Poisson point process . . . 160
- 127. Tails of exponentials . 160
- 129. Bernoullis from geometrics 160
- 131. Understanding Poisson point processes . . . 160
- 133. Conditional PPP . . . 160
- 135. The restaurant 161

Densities for continuous random variables 161
- 137. Density of a uniform . 161
- 139. More uniform density 161
- 141. Another exponential density 161
- 143. Normalizing an exponential 162
- 145. Density of a function of a uniform 162
- 147. From cdf to pdf 162

Densities for discrete random variables 162
- 149. Pdf for a die 162
- 151. Binomial mode 162
- 153. Gamma median . . . 163
- 155. Cdf of a scaled die . . 163
- 157. Scaling a beta 163
- 159. Survival function of an exponential . . . 163

Mean of a random variable . . . 164
- 161. Mean of finite random variables 164
- 163. Mean of a discrete uniform 164
- 165. Mean of a die roll . . 164
- 167. Mean of a difference . 164
- 169. Symmetry of a discrete uniform 164
- 171. Vertigon's Army . . . 164
- 173. Four stores 165
- 175. The SLLN in action . 165
- 177. Two tasks 165

Mean of a general random variable 165
- 179. Mean of an exponential 165
- 181. Mean of functions of a continuous uniform 165
- 183. Monte Carlo with uniforms 166
- 185. Polynomials of continuous uniforms 166
- 187. A nonintegrable density 166
- 189. Deriving formulas . . 166

Conditional expectation 166
- 191. Raising sales 166
- 193. Food for thought . . . 167
- 195. Random numbers of dice 167
- 197. The Bayesian, Part I . 168
- 199. Using the FTP 168

Joint densities of random variables 168
- 201. Joint densities 168
- 203. Independence 169
- 205. Multiple integrals . . 169
- 207. Discrete joint densities 170
- 209. Factoring continuous densities 170
- 211. Dependent joint continuous uniforms 170

Random variables as vectors . . 171
- 213. Rules of inner products 171
- 215. Rules of inner product norms 171
- 217. Standard deviation of discrete random variables 171
- 219. Covariance of a joint uniform 171

Correlation 172
- 223. Affine transforms . . 172

- 225. Correlation of discrete uniforms 172
- 227. Understanding functions 173
- 229. The Pythagorean Theorem 173
- 231. Sample averages ... 173

Adding random variables 173
- 233. Adding random variables 173
- 235. Adding independent dice 174
- 237. Adding continuous random variables .. 174
- 239. Generating function of a Poisson 174
- 241. Wolfram Alpha to the rescue 175

The Central Limit Theorem ... 175
- 243. CLT for exponentials 175
- 245. CLT for geometrics .. 175
- 247. Another exponential CLT 175

Normal random variables 176
- 253. Shifting and scaling normals 176
- 255. Adding independent normals 176
- 257. Multivariate scaling and shifting 176

Bayes rule for densities 176
- 259. Archytas Medical ... 176
- 261. Toothpaste troubles . 176
- 263. Ronco Survey Group 177
- 265. Dimer Medicine ... 177
- 267. The assembly line .. 178
- 269. Rush Hour 178

The multinomial distribution . 179
- 271. Probabilities for multinomials 179
- 273. Means of multinomials 179
- 275. Multinomial covariance 179

Tail inequalities 180
- 277. Markov's inequality . 180
- 279. Chebyshev's inequality 180
- 281. Chebyshev via standard deviation 180
- 283. Sample averages and Chebyshev 180
- 285. Construction woes .. 180

Chernoff inequalities 181
- 287. Chernoff for a Poisson 181
- 289. Chernoff for gammas 181
- 291. Chernoff given the mgf 181
- 293. Chernoff for uniforms 181

The hypergeometric distribution 182
- 295. Those crazy eights .. 182
- 297. A bag of marbles ... 182
- 299. A box of screws 182
- 301. The mission 182
- 303. The forest 183

More Poisson point processes . 183
- 305. Back to the cellar .. 183
- 307. The restaurant 184
- 309. An abstract space .. 184

Transforming multivariate random variables 184
- 311. One dimensional transform 184
- 313. Two dimensional transform 184

Chapter 1: What is Probability?

HE RANGER WAS SENT OUT TO scout an invading horde of monsters. There were three possibilities: the horde was either large, huge, or massive. Given that each of these possibilities were equally likely, what was the chance that the horde was massive?

Probability & Logic

In logic, statements are either *true* or *false*, and there is no in between. To make things numerical, a true statement can be assigned a value of 1, and a false statement a value of 0. But in reality, things are not always so clear cut. Instead, it is helpful to have numbers between 0 and 1 that represent the *information* that is available about the truth of the statement.

So a *probability* that a statement is true could be 10%, or 65%, or $1/e$, or any other number from 0 to 1. The statement itself is either true or false, what the probability represents is the information about the two possibilities.

True and False

The idea of *true* and *false* will come up often enough that it is good to have notation for them. In this text, T will stand for true, and F will stand for false. Anything in mathematics that evaluates to be either T or F is called a logical statement.

D1 A **logical statement** is any statement that is either T or F.

The statement was preceded by **D1** to indicate that this was our first definition. Other special statements will be preceded by **F** for facts, **T** for theorems, and **E** for examples.

Indicator functions

A *function* can be defined more formally, but it is helpful to think of it as an operation that transforms values into other values. One of the simplest functions is the *indicator function*. The indicator function (often denoted $\mathbb{I}$), transforms T into a 1 and F into a 0.

D2 The **indicator function** is defined as
$$\mathbb{I}(T) = 1$$
$$\mathbb{I}(F) = 0.$$

E1 What is $\mathbb{I}(3 < 5)$? What is $\mathbb{I}(3 < 1)$?
Answer. The statement $(3 < 5)$ is true, so $\mathbb{I}(3 < 5) = \boxed{1}$. The statement $(3 < 1)$ is false, so $\mathbb{I}(3 < 1) = \boxed{0}$.

E2 What is $\mathbb{I}(x^2 + 5 > 4)$?
Answer. No matter what x is, $(x^2 \geq 0)$ and $(x^2 + 5 > 4)$, so $\mathbb{I}(x^2 + 5 > 4) = \boxed{1}$.

E3 Suppose $g(x) = \mathbb{I}(x^2 > 4)$.
  a. Evaluate $g(3)$.
  b. Evaluate $g(1)$.
  c. Evaluate $g(-3)$.
Answer. These can be found as follows.
  a. Since $3^2 = 9 > 4$, $g(3) = \mathbb{I}(9 > 4) = \boxed{1}$.
  b. Here $g(1) = \mathbb{I}(1 > 4) = \boxed{0}$.
  c. Here $g(-3) = \mathbb{I}((-3)^2 > 4) = \boxed{1}$.

For example $\mathbb{I}(3 < 5) = 1$ and $\mathbb{I}(3 < 1) = 0$ since the statement $3 < 5$ is true but $3 < 1$ is false.

A nice thing that will come in handy later, is that if there are multiple logical statements involved, indicator functions can be used to count how many of the statements are true. For instance,

$$\mathbb{I}(3<5)+\mathbb{I}(3<1)+\mathbb{I}(7<2)=1+0+0=1,$$

and exactly one of the statements

$$(3<5), (3<1), (7<2)$$

is true.

This fact can be written more generally.

> **F1** If $s_1, \ldots, s_n$ are a finite set of statements, then
> $$\mathbb{I}(s_1) + \cdots + \mathbb{I}(s_n)$$
> counts how many of the statements are true. That is,
> $$\#(i : s_i = \mathsf{T}) = \sum_{i=1}^{n} \mathbb{I}(s_i).$$

A fact such as this needs a proof so that a user can be sure that it is correct.

Proof. Use mathematical induction on the number of statements n. For the base case of $n = 1$, if $s_1 = \mathsf{T}$ then $\mathbb{I}(s_1) = 1$, and if $s_1 = \mathsf{F}$ then $\mathbb{I}(s_1) = 0$. Either way $\mathbb{I}(s_1)$ counts the number of true statements in the set.

The induction hypothesis is that the fact holds for n statements. Now consider statements $s_1, \ldots, s_{n+1}$. Since $\{1, \ldots, n\}$ and $\{n+1\}$ are disjoint sets (they have no elements in common), the number of statements that are true equals the number of statements s_i for i from 1 to n that are true, with a 1 added if and only if the statement s_{n+1} is true.

By the induction hypothesis, the number of $s_1, \ldots, s_n$ that are true is $\sum_{i=1}^{n} \mathbb{I}(s_i)$, and then adding $\mathbb{I}(s_{n+1})$ adds 1 to this total if and only if s_{n+1} is true. Hence the total is

$$\left[\sum_{i=1}^{n} \mathbb{I}(s_i)\right] + \mathbb{I}(s_{n+1}) = \sum_{i=1}^{n+1} \mathbb{I}(s_i),$$

and the induction is complete. □

> **E4** If $\sum_{i=1}^{10} \mathbb{I}(b_i) = 7$, how many of $b_1, \ldots, b_{10}$ are true?
> **Answer.** Since the sum of the indicators is 7, exactly $\boxed{7}$ must be true.

In the story of the Ranger from the beginning of the chapter, the size of the horde could be either large, huge, or massive. The events large, huge, massive are *disjoint* or *mutually exclusive*, meaning that at most one of the possibilities can be true at any one time. Using indicator function, this idea can be more clearly expressed.

> **D3** A set of statements $\{s_1, \ldots, s_n\}$ are disjoint (aka mutually exclusive) if
> $$\sum_{i=1}^{n} \mathbb{I}(s_i) \leq 1.$$

PROBABILITY FUNCTIONS

While indicator functions are great if the truth or falsehood of a statement is known precisely, there are many instances where it is not. For example, something might not have happened yet, like the next Presidential election in the United States.

Or the statement could be about a physical system where small changes in the initial values lead to big changes over time. Flipping coins and rolling dice fall into this category.

Another possibility is that the information is too expensive to obtain exactly. For instance, determining the exact number of people living in a major city can be very difficult to obtain.

A *probability function* assigns some logical statements a number from 0 up to 1 that indicates the knowledge about the truth of the statement. Usually $\mathbb{P}$ will be used to denote this function. As with indicator functions, $\mathbb{P}(\mathsf{T}) = \mathbb{I}(\mathsf{T}) = 1$, and $\mathbb{P}(\mathsf{F}) = \mathbb{I}(\mathsf{F}) = 0$, but a probability function can also have $\mathbb{P}(s_1) = 0.3$ or $\mathbb{P}(s_2) = 5/17$ for statements s_1 and s_2.

> **E5** What is $\mathbb{P}(x^2 + 5 > 4)$?
> **Answer.** No matter what x is, $(x^2 \geq 0)$ and $(x^2 + 5 > 4)$, so $\mathbb{I}(x^2 + 5 > 4) = \boxed{1}$.

When it is possible to assign a probability to a logical statement, call the statement an *event*.

> **D4** If $\mathbb{P}(s)$ exists for a logical statement s, then s is an **event**.

ASSIGNING PROBABILITIES

In the Ranger's story, each of the three possibilities of large, huge, and massive for the size of the horde were equally likely. If the total amount of truth available is 1, and there are three possibilities, then the simplest way to divide the truth is evenly among the three.

In other words, each of the three possibilities should have $1/3$ probability. In general, this idea is known as the *Principle of Indifference*.

> **D5** The **Principle of Indifference** says that if exactly one of $s_1, \ldots, s_n$ must be true, and there is no reason to believe that any one event is more likely than another, then for each i, $\mathbb{P}(s_i) = 1/n$.

THE PRINCIPLE OF INDIFFERENCE
This was developed as an idea by Jakob Bernoulli and Pierre-Simon Laplace in the 1800's. It was given the name Principle of Indifference by the 20th century economist John Maynard Keynes.

MATHEMATICAL PROBABILITY

The events that are assigned probabilities often involve *sets*, which are collections of objects called *elements*.

SETS

If an object a is in the set S then write $(a \in S) = \mathsf{T}$ and call a an element of S. Otherwise $(a \in S) = \mathsf{F}$, also written $(a \notin S) = \mathsf{T}$, and a is not an element of S.

> **D6** Say that S is a **set** if for any object a, the expression
> $$a \in S$$
> is a logical statement that is either T or F. If $(a \in S) = \mathsf{T}$, then a is an **element** of S.

SET TERMINOLOGY
Another way to read $a \in S$ aloud is to say that S **contains** the element a. Another term for a set is a **collection**.

For instance, suppose that the set S has
$$(\text{red} \in S) = \mathsf{T}$$
$$(\text{green} \in S) = \mathsf{T},$$
and no other objects than red or green are elements of the set. Note that there is no notion of an ordering of the objects red or green, the two equations written in reverse order would describe exactly the same set.

A shorter way to write this set is to set *set notation*, where the elements of the set are listed out and surrounded by curly braces. So the same set could be written as
$$S = \{\text{red}, \text{green}\}.$$

A special set is the *positive integers*. This set is usually denoted by
$$\{1, 2, 3, \ldots\}.$$

CHAPTER 1: WHAT IS PROBABILITY?

COINS AND DICE

A simple probability experiment comes from flipping a fair coin. A coin typically has two sides, one labeled heads for convenience, and the other side labeled tails. If the coin is fair, then the Principle of Indifference holds, and each side of the coin is equally likely to come up when flipped.

Figure 1.2: Bone dice excavated from Syria (https://artgallery.yale.edu/collections/objects/6521).

Figure 1.1: Copper coin minted in 1787 (https://artgallery.yale.edu/collections/objects/170864).

Another simple probability experiment is to roll a fair die. Die is the singular form, and dice is the plural. One common type of die has six sides, each of which is likely to show on top when rolled.

The notation d followed by a number indicates the outcome of a roll of a fair die with that number of sides. For instance, d6 indicates a roll of a die with sides labeled 1 through 6. So if X was the outcome of the roll, it is always true that $X \in \{1, 2, 3, 4, 5, 6\}$, which can be written in shorthand as $X \in \{1, 2, \ldots, 6\}$.

Since this is always true, and true statements have probability 1, $\mathbb{P}(X \in \{1, 2, \ldots, 6\}) = 1$. Because the die is fair, the Principle of Indifference applies, and $\mathbb{P}(X = 4) = 1/6$ since there are six sides to the die.

Write $X \sim$ d6 to indicate that the value of X comes from the roll of the six sided die.

E6 Suppose $X \sim$ d6 and $Y \sim$ d8. If the Principle of Indifference holds, what is $\mathbb{P}((X, Y) = (3, 2))$?

Answer. There are six choices for X and eight choices for Y. Hence there are $6 \cdot 8 = 48$ choices for (X, Y). That means that (assuming the Principle of Indifference holds)

$$\mathbb{P}((X, Y) = (3, 2)) = 1/48.$$

This is about $\boxed{0.02083}$.

ENCOUNTERS

1. BASIC INDICATORS

What is $\mathbb{I}(4^2 > 10)$?

2. MORE BASIC INDICATORS

For x a real number, what is $\mathbb{I}(x^2 \geq 0)$?

3. BASIC PROBABILITIES

What is $\mathbb{P}(4^2 > 10)$?

4. MORE BASIC PROBABILITIES

For x a real number, what is $\mathbb{P}(|x| \geq 0)$?

5. Indicator functions

Suppose $f(x) = \mathbb{I}(|x| > 4)$.
a. What is $f(2)$?
b. What is $f(-2)$?
c. What is $f(5)$?
d. What is $f(-5)$?

6. Indicator function factors

Suppose $g(x) = \exp(-x)\mathbb{I}(x \geq 0)$.
a. What is $g(2)$?
b. What is $g(-2)$?
c. What is $g(0)$?

7. Graphing indicator functions

Indicator functions, when used to make functions, can be graphed.
a. Suppose $f(x) = \mathbb{I}(|x| > 4)$. Graph $f(x)$.
b. Suppose $g(x) = (|x|/4)\mathbb{I}(|x| > 4)$. Graph $g(x)$.

8. Graphing indication function factors

Graph $h(x) = \exp(-x)\mathbb{I}(x \geq 0)$.

9. Disjoint reals

Suppose X is a real number. State if the following events are disjoint or not.
a. $(X \leq 3)$ and $(X \geq 4)$.
b. $(X \leq 5)$ and $(X \geq 3)$.
c. $(X \leq 3)$ and F.

10. Disjoint integers

Suppose N is an integer. State if the following events are disjoint or not.
a. $(N = 3)$ and $(N = 4)$.
b. $(N = 3)$ and $(N \geq 3)$.
c. $(N = 3)$ and $(N = 4)$ and $(N = 5)$.

11. Truth or dare?

What is $\mathbb{P}(10 < 20)$?

12. Truth or dare, arithmetic edition

What is $\mathbb{P}(3 + 7 > 5)$?

13. Counting true statements

Suppose s_1, s_2, and s_3 are disjoint statements. What is the largest that $\mathbb{I}(s_1) + \mathbb{I}(s_2) + \mathbb{I}(s_3)$ can be?

14. Adding indicators

If $\sum_{i=1}^{5} \mathbb{I}(a_i) = 4$, how many of $a_1, a_2, \ldots, a_5$ are true?

15. Two dice

If A is the roll of a fair six-sided die, (so $A \sim$ d6) and B is the roll of a fair four-sided die (so $B \sim$ d4), how many different outcomes can there be for (A, B)?

16. Three dice

If a six-sided die is rolled three times to give (D_1, D_2, D_3), how many possible combinations are there?

17. Principle of indifference

If A is the roll of a fair six-sided die, (so $A \sim$ d6) and B is the roll of the fair four-sided die (so $B \sim$ d4), what would $\mathbb{P}((A, B) = (3, 1))$ using the Principle of Indifference?

18. Principle of indifference for three dice

If three fair six-sided dice are rolled, what is the chance that all three rolls are 3 using the Principle of Indifference?

Chapter 2: Logical Operators

The Ranger studying the horde of monsters believes that it either contains a hippogriff or a bugbear or both. What is the negation of that statement?

Modifying Logical Statements

When calculating probabilities, it is often useful to be able to tie together or break apart logical statements into their pieces. In order to accomplish this, it is necessary to understand the main ways in which logical statements can be connected.

Logical OR

The *logical OR* operator is applied to a pair of logical statements, and returns true when either one or both of the statements are true. The logical OR of p and q is written $p \vee q$. For instance,

$$(3 = 3) \vee (3 < 4) = T \vee T = T$$
$$(3 = 3) \vee (4 < 3) = T \vee F = T$$
$$(3 < 4) \vee (6 = 8) = T \vee F = T$$
$$(4 < 3) \vee (6 = 8) = F \vee F = F$$

This is equivalent to saying that the logical OR is true when at least 1 of the statements is true. This leads to the following definition in terms of indicator functions.

D7 The **logical OR** between statements s_1 and s_2 is

$$(s_1 \vee s_2) = (\mathbb{I}(s_1) + \mathbb{I}(s_2) \geq 1).$$

English versus CS
Computer scientists often capitalize the OR in logical OR to distinguish it from the common use of the word *or* in English. In English, or often means that one or the other but not both are true. In logical OR if both the statements are true, then the logical OR is also true.

E7 State if the following are true or false:
a. $(10 = 10) \vee (3 < 4)$.
b. $(10 = 11) \vee (3 < 4)$.
c. $(10 = 11) \vee (4 < 3)$.

Answer. Consider the terms in the expressions.
a. Since $(10 = 10) = T$, the overall statement is $\boxed{T}$.
b. Since $(3 < 4) = T$, the overall statement is $\boxed{T}$.
c. Since both $(10 = 11)$ and $(4 < 3)$ are false, the overall statement is $\boxed{F}$.

Since logical OR was defined using addition, which is commutative, so is logical OR.

F2 For logical statements s_1 and s_2,

$$s_1 \vee s_2 = s_2 \vee s_1.$$

This definition can be extended nicely to any finite number of statements $s_1, \ldots, s_n$, or even to a countable sequence of statements.

D8 The **countable logical OR** of $s_1, s_2, \ldots$ is
$$\left(\bigvee_{i=1}^{\infty} s_i\right) = \left(\sum_{i=1}^{\infty} \mathbb{I}(s_i) > 0\right).$$

Logical AND

The logical AND of two statements s_1 and s_2, written $s_1 \wedge s_2$, is true if and only if both of the statements are true. For instance,

$$(3 = 3) \wedge (3 < 4) = \mathsf{T}$$
$$(3 = 3) \wedge (4 < 3) = \mathsf{F}$$
$$(3 < 4) \wedge (6 = 8) = \mathsf{F}$$
$$(4 < 3) \wedge (6 = 8) = \mathsf{F}$$

Logical OR corresponds to addition, since
$$\bigvee s_i = \left(\sum \mathbb{I}(s_i) > 0\right),$$
whereas logical AND is more like multiplication, since
$$\bigwedge s_i = \left(\prod \mathbb{I}(s_i) > 0\right).$$

D9 The **logical AND** of $s_1, s_2, \ldots, s_n$ is
$$\left(\bigwedge_{i=1}^{n} s_i\right) = \left(\prod_{i=1}^{n} \mathbb{I}(s_i) > 0\right).$$

The **countable logical AND** of $s_1, s_2, \ldots$ is
$$\left(\bigwedge_{i=1}^{\infty} s_i\right) = \left(\prod_{i=1}^{\infty} \mathbb{I}(s_i) > 0\right).$$

Notation for logical AND
Alternative notation for logical AND includes placing a comma between the statements, or just pushing the statements together (as with multiplication.)

$$(s_1 \wedge s_2) = (s_1, s_2) = (s_1 s_2).$$

E8 State if the following are true or false:
a. $(10 = 10) \vee (3 < 4)$,
b. $(10 = 11) \vee (3 < 4)$,
c. $(10 = 11) \vee (4 < 3)$.

Answer. These evaluate as follows.
a. Since $(10 = 10) = \mathsf{T}$, the overall statement is $\boxed{\mathsf{T}}$.
b. Since $(3 < 4) = \mathsf{T}$, the overall statement is $\boxed{\mathsf{T}}$.
c. Since both $(10 = 11)$ and $(4 < 3)$ are false, the overall statement is $\boxed{\mathsf{F}}$.

Logical NOT

The third major logical operator is logical NOT, which is also called *negation*. What it does is change a T to a F and a F to a T. It can be defined (as with the other operators) using indicator functions.

D10 The **logical NOT** (aka **negation**) of s is
$$(\neg s) = (\mathbb{I}(s) = 0).$$

Negation can be used to define the *complement* of a set, the set of elements not in the original set.

D11 The **complement** of a set A, written A^C is defined as
$$(x \in A^C) = \neg(x \in A).$$

Order of Operations

When adding and multiplying numbers, there is an order of operations. Multiplication is done before addition. For instance,

$$3 + 4 \cdot 6 = 3 + 24 = 27,$$

because the multiplication is done before the addition in the statement.

In the same way, there is an order of operations of logical operators. First comes logical NOT, then logical AND, and finally logical OR. That is,

NOT before AND before OR

E9 State if the following are true or false.
 a. $\neg(10 = 10)$,
 b. $\neg(10 = 11)$,
 c. $\neg(10 = 11) \wedge (3 < 4)$.
 d. $\neg(10 = 11) \vee (5 > 6) \wedge (3 < 4)$.

Answer. First apply the negation, then other operators.
 a. $\neg(10 = 10) = \neg T = \boxed{F}$.
 b. $\neg(10 = 11) = \neg F = \boxed{T}$.
 c. $\neg(10 = 11) \wedge (3 < 4) = \neg F \wedge T = T \wedge T = \boxed{T}$.
 d. $\neg(10 = 11) \vee (5 > 6) \wedge (3 < 4) = \neg F \vee F \wedge T = T \vee F = \boxed{T}$.

LOGIC AND ARITHMETIC

Fortunately, there are three facts about indicator functions that enable us to turn problems involving logic into multiplication and addition.

F3 For logical statements s and r, the following three statements hold.

$$\mathbb{I}(\neg s) = 1 - \mathbb{I}(s) \quad \text{Negation}$$
$$\mathbb{I}(s \vee r) = \mathbb{I}(s) + \mathbb{I}(r) - \mathbb{I}(s)\mathbb{I}(r) \quad \text{logical OR}$$
$$\mathbb{I}(sr) = \mathbb{I}(s)\mathbb{I}(r) \quad \text{logial AND}.$$

These three facts about indicators allow us to break apart any logical statement involving a finite number of NOT, AND, and OR into additive and multiplicative statements about the indicator function of the individual variables.

E10 Write $\mathbb{I}((s \vee \neg r) \wedge t)$ as an expression that only uses constants, $\mathbb{I}(s)$, $\mathbb{I}(r)$, and $\mathbb{I}(t)$.

Answer. First use the logical AND rule to write

$$\mathbb{I}((s \vee \neg r) \wedge t) = \mathbb{I}((s \vee \neg r))\mathbb{I}(t).$$

Next, the logical OR rule gives us

$$\mathbb{I}((s \vee \neg r))\mathbb{I}(t) = [\mathbb{I}(s) + \mathbb{I}(\neg r) - \mathbb{I}(s)\mathbb{I}(\neg r)]\mathbb{I}(t).$$

The negation rule then gives

$$\mathbb{I}((s \vee \neg r))\mathbb{I}(t) = [\mathbb{I}(s) + 1 - \mathbb{I}(r) - \mathbb{I}(s)(1 - \mathbb{I}(r))]\mathbb{I}(t).$$

Finally, simplify to obtain

$$\mathbb{I}((s \vee \neg r))\mathbb{I}(t) = \boxed{[1 - \mathbb{I}(r) + \mathbb{I}(s)\mathbb{I}(r)]\mathbb{I}(t)}.$$

Proof. The easiest way to verify that these three rules hold is with a *truth table*. Such a table lists out the possible values for the variables, and checks that the equations are equal. For instance,

NEGATION RULE TRUTH TABLE

s	$\neg s$	$\mathbb{I}(\neg s)$	$1 - \mathbb{I}(s)$
T	F	0	$1 - 1 = 0$
F	T	1	$1 - 0 = 1$

So not matter what the value of s, the Negation rule holds!

For the logical AND rule, there are two choices for s and two for r, leading to four possible choices for (s, r).

LOGICAL AND RULE TRUTH TABLE

(s, r)	$\mathbb{I}(sr)$	$\mathbb{I}(s)\mathbb{I}(r)$
(T, T)	1	$1 \cdot 1 = 1$
(T, F)	0	$1 \cdot 0 = 0$
(F, T)	0	$0 \cdot 1 = 0$
(F, F)	0	$0 \cdot 0 = 0$

Since the second and third columns are equal, the logical AND rule holds.

Finally, the truth table for logical OR.

LOGICAL OR RULE TRUTH TABLE

(s, r)	$\mathbb{I}(s \vee r)$	$\mathbb{I}(s) + \mathbb{I}(r) - \mathbb{I}(s)\mathbb{I}(r)$
(T, T)	1	$1 + 1 - 1 = 1$
(T, F)	1	$1 + 0 - 0 = 1$
(F, T)	1	$0 + 1 - 0 = 1$
(F, F)	0	$0 + 0 - 0 = 0$

□

The indicator function is a one-to-one function, which immediately gives the following.

F4 For logical statements p and q,
$$(p = q) = (\mathbb{I}(p) = \mathbb{I}(q)).$$

This allows us to work out more complicated equations using arithmetic instead of logic.

F5 For a logical statement s,
$$\neg\neg s = s.$$

Proof. Note
$$\mathbb{I}(\neg\neg s) = 1 - \mathbb{I}(\neg s) = 1 - (1 - \mathbb{I}(s)) = \mathbb{I}(s).$$

Since the two statements have the same indicator function, they have the same truth value. □

Indicators over logical AND become multiplication.

F6 For a sequence of logical statements $s_1, s_2, \ldots$,
$$\mathbb{I}(s_1 s_2 \cdots s_n) = \mathbb{I}(s_1)\mathbb{I}(s_2)\cdots\mathbb{I}(s_n)$$
$$\mathbb{I}(s_1 s_2 \cdots) = \mathbb{I}(s_1)\mathbb{I}(s_2)\cdots.$$

Proof. The first statement follows by induction using $\mathbb{I}(s_1 s_2) = \mathbb{I}(s_1)\mathbb{I}(s_2)$.

For the countable sequence, by definition
$$\mathbb{I}(s_1 s_2 \cdots) = \mathbb{I}\left(\prod_{i=1}^{\infty} \mathbb{I}(s_i) > 0\right).$$

That is 1 if and only if the infinite product is 1, which in turn means that for any finite n, $\prod_{i=1}^{n} \mathbb{I}(s_i) = 1$. The limit as n goes to infinity then gives the right hand side. □

De Morgan's Laws

At this point something more advanced can be proved, namely, De Morgan's Laws. These allow logical NOT to be distributed over logical OR and logical AND. First, consider distribution over logical OR.

F7 For events s and r,
$$\neg(s \vee r) = \neg s \wedge \neg r.$$

Proof. Compare indicator functions:
$$\mathbb{I}(\neg(s \vee r)) = 1 - \mathbb{I}(s \vee r)$$
$$= 1 - [\mathbb{I}(s) + \mathbb{I}(r) - \mathbb{I}(sr)]$$
$$= 1 - \mathbb{I}(s) - \mathbb{I}(r) + \mathbb{I}(s)\mathbb{I}(r)$$
$$= (1 - \mathbb{I}(s))(1 - \mathbb{I}(r))$$
$$= \mathbb{I}(\neg s)\mathbb{I}(\neg r)$$
$$= \mathbb{I}(\neg s \wedge \neg r).$$
□

For example, suppose y is the logical statement that a ball is yellow, and b is the logical statement that a ball is blue. Then
$$\neg(y \vee b)$$
is the logical statement that the ball is neither yellow nor is it blue.

The logical statement
$$\neg y \wedge \neg b$$
is the statement that the ball is not yellow and not blue.

De Morgan's law states that these two statements are the same. These laws can be generalized as follows.

T1 De Morgan's Laws For a sequence $s_1, s_2, \ldots$ of events, and $n = \in \{1, 2, \ldots\}$, De Morgan's laws state
$$\neg \bigvee s_i = \bigwedge \neg s_i$$
$$\neg \bigwedge s_i = \bigvee \neg s_i.$$

Proof. To see the first result, suppose at least one of the s_i is true. Then the logical OR is true, and the negation is

CHAPTER 2: LOGICAL OPERATORS

false. On the right hand side, at least one of the $\neg s_i$ is false, making the logical AND false.

The second result is similar. □

E11 To write

$$\neg(s_1 \lor s_2 \lor \neg s_3)$$

using only logical AND and logical NOT, apply De Morgan's Laws to state

$$\neg(s_1 \lor s_2 \lor \neg s_3) = \neg s_1 \land \neg s_2 \land s_3.$$

SOLVING THE STORY

In the Ranger's story at the beginning, let

$s_1 =$ The horde contains a hippogriff
$s_2 =$ The horde contains a bugbear

Then the Ranger believes that $s_1 \lor s_2$ is true. The question to be considered is what is the negation of this event?

Using De Morgan's Laws:

$$\neg(s_1 \lor s_2) = \neg s_1 \land \neg s_2.$$

In English, the negative of the statement "the horde contains either a hippogriff or a bugbear (or both)" is "the horde does not contain a hippogriff and it does not contain a bugbear."

CIRCUITS
All digital computers are created using *circuits*, which employ either logical OR, logical AND, or logical NOT to solve problems. Any circuit can be broken down into these component parts. In the same way, our methods of calculating probabilities will break apart events into pieces using logical OR, logical AND, and logical NOT.

IMPLICATION

Say that s_1 *implies* s_2 if whenever s_1 is true, then s_2 must be true as well. For instance, $(x > 3)$ implies that $(x > 0)$, since whenever $(x > 3)$ is true, it must also be true that $(x > 0)$. Note that the reverse implication does not hold: just because $(x > 0)$, there is no guarantee that $(x > 3)$ will be true.

In the end, this means that either $\neg p$ is true or p and q are both true. Using the distribution laws:

$$\neg p \lor (p \land q) = (\neg p \lor p) \land (\neg p \lor q)$$
$$= \mathsf{T} \land (\neg p \lor q)$$
$$= (\neg p \lor q).$$

D12 Say that s_1 implies s_2 (write $s_1 \to s_2$) to mean

$$(s_1 \to s_2) = (\neg s_1 \lor s_2).$$

SUBSETS

Logical implication is closely tied to the notion of one set being a *subset* of another set.

D13 For two sets A and B, say that A is a *subset* of B and write $A \subseteq B$ to mean

$$(A \subseteq B) = ((x \in A) \to (x \in B)).$$

For instance, $\{1, 5, 7\} \subseteq \{1, 2, \ldots, 7\}$ since

$$(x \in \{1, 5, 7\}) \to (x \in \{1, \ldots, 7\}).$$

ENCOUNTERS
19. LOGICAL AND, OR, AND NOT

State if the following are true or false.
a. $(1 < 6) \lor (6 < 1)$
b. $(1 < 6) \land (6 < 1)$
a. $(6 < 1) \lor (1 < 6) \land (10 > 20)$

20. NEGATION

State if the following logical statements are true or false.
a. $\neg(1 > 6)$

b. $\neg(6 > 1) \lor (7 > 3)$
c. $\neg(6 > 1) \lor (7 > 3) \land (4 > 2)$

21. Countable logical AND

Let x be a a positive integer (so $x \in \{1, 2, 3, \ldots\}$ is true.) State if $(x \geq 1) \land (x \geq 2) \land (x \geq 3) \land \cdots$ is true or false.

22. Countable logical OR

Let x be a a positive integer (so $x \in \{1, 2, 3, \ldots\}$ is true.) State if $(x \geq 1) \lor (x \geq 2) \lor (x \geq 3) \lor \cdots$ is true or false.

23. Logic to arithmetic

Write $\mathbb{I}(\neg s \lor \neg r)$ using only constants, $\mathbb{I}(s)$, and $\mathbb{I}(r)$.

24. More logic to arithmetic

Write $\mathbb{I}(\neg(s \land r))$ using only constants, $\mathbb{I}(s)$, and $\mathbb{I}(r)$.

25. Proof with indicator functions

Prove that $\neg(s \land r) = \neg s \lor \neg r$ using indicator functions.

26. More proof with indicator functions

Show that $\neg(p \lor q) = \neg p \land \neg q$ by comparing their indicator functions.

27. Logical order of operations

Write the order of operations for $s \lor w \land \neg r$ explicitly using parentheses.

28. More logical order of operations

Write the order of operations for the following logical expression using parentheses.

$$\neg a_1 \lor a_2 \land a_3 \lor a_4.$$

29. De Morgan's Laws

Write $\neg(s_1 \land \neg s_2)$ using only logical OR and logical NOT.

30. More De Morgan's Laws

Write $\neg(\neg s_1 \lor s_2 \lor \neg s_3)$ using only logical AND and logical NOT.

31. Implication and subsets

State whether the following are true or false.
a. $(x > 4) \to (x > 10)$.
b. $(x > 4) \to (x > 3)$.
c. $\{a, b, c\} \subseteq \{a, b, c, d\}$.
d. $\{a, b, c\} \subseteq \{a, c, e, g\}$.

32. More implications and subsets

State if the following are true or false.
a. $(x = 3) \to (x < 4)$.
b. $(x = 3) \to (x < 3)$.
c. $\{2, 4, 6, \ldots\} \subseteq \{1, 2, 3, \ldots\}$.
d. $\{1, 2, 3, \ldots\} \subseteq \{2, 4, 6, \ldots\}$.

Chapter 3: The Rules of Probability

A Giant Problem. As the party of adventures ran from the group of monsters, they realized that there was a 60% chance the monsters contained a Giant, 70% chance that the monsters contained an Ogre, and a 40% chance that the horde contained both. What was the chance that the horde contained either a Giant, an Ogre, or both?

Mathematical probability

A *probability function* takes as input an event that might be true or false, and returns a number from 0 up to 1 that gives the information about the truth of the event. Now consider how to make this notion precise.

The collection of events

An *event* is just any logical statement that can be assigned a probability. Let $\mathcal{F}$ denote the collection of all events. The goal is to make $\mathcal{F}$ the domain of our probability function. That is to say, the elements of $\mathcal{F}$ will be the possible inputs to the probability function. It turns out that it will be helpful that $\mathcal{F}$ has several properties.

First, any event that is always true will be assigned probability 1. Hence $\mathsf{T} \in \mathcal{F}$.

Second, if an event is in $\mathcal{F}$ that means it is possible to assign a probability that the event is true. That in turn means it should be possible to assign a probability that the event is false. In other words, if $s \in \mathcal{F}$, then $\neg s \in \mathcal{F}$ should also hold.

The third (and last) property is a bit more unusual. Suppose a fair coin with outcomes heads and tails is flipped over and over again and s_i is the event that the first $i - 1$ flips were tails and the ith flip of the coin was heads. So if the first few coin flips were

$$TTHTH\ldots$$

then $s_1 = s_2 = \mathsf{F}$, while $s_3 = \mathsf{T}$. The rest of the events $s_4 = s_5 = s_6 = \ldots$ are all false with this sequence of coin flips.

Consider the logical statement

$$s_1 \vee s_2 \vee \cdots.$$

This is the logical statement that the number of flips until the first head is either 1, or 2, or 3, and so on. Another way to say this is the event that the number of flips needed for the first head is some finite integer. Given that each s_i is assigned a probability, it would be useful to also guarantee that their logical OR was also an event, that is, that it could also be assigned a probability.

Therefore, our last property is that if there are a countable sequence of events then the logical OR of that sequence is also an event. If these three properties hold for the collection of events, call the collection a σ-algebra.

D14 A nonempty set $\mathcal{F}$ is a σ-**algebra** if for all sequences $s_1, s_2, \ldots$ of events in $\mathcal{F}$, the following three properties hold.
1. $\mathsf{T} \in \mathcal{F}$.
2. $\neg s_1 \in \mathcal{F}$.
3. $s_1 \vee s_2 \vee \cdots \in \mathcal{F}$.

NAMING σ-ALGEBRAS
Why do we usually use a script capital letter $\mathcal{F}$ for a σ-algebra? Because the French word for *closed* is *fermé*, and the collection of sets is closed under negation and logical OR.

THE PROBABILITY FUNCTION

Now that the properties of the set of events is clear, it is possible to formally define what is a probability function. There are two ideas. The first is that the probability of a true statement should be 1. That is just so that probability functions are extensions of indicator functions.

The next part is trickier, so it helps to return to our example of an infinite sequence of coin flips. Suppose that N represents the number of flips needed to get a head on the coin. So $N = 1$ means the very first flip was a head. And $N = 2$ means the first flip was a tail, and then the second flip was a head, and so on.

Note that $N = 1$ and $N = 2$ are *disjoint events* meaning that they cannot both happen at the same time. That is, they cannot both be true simultaneously. So the probability that either of those two events occurring should just be the sum of the probability of the two events. That is, if $\mathbb{P}(N=1) = 0.3$ and $\mathbb{P}(N=2) = 0.21$, then

$$\mathbb{P}(N=1 \vee N=2) = \mathbb{P}(N \in \{1, 2\})$$
$$= 0.3 + 0.21$$
$$= 0.51.$$

In fact,

$$(N=1), (N=2), (N=3), \ldots$$

form an entire sequence of disjoint events because at most one can be true at a time. Just like in the example with two disjoint events, the probability that any of those sequence of disjoint events occurring should just be the sum of the individual probabilities that the events occur.

When that holds for all disjoint sequences of events (together with the rule that $\mathbb{P}(\mathsf{T}) = 1$), the result is a *probability function*.

D15 Probability Function
For $\mathcal{F}$ a σ-algebra, $\mathbb{P} : \mathcal{F} \to [0,1]$ is a **probability function** (aka **probability measure** aka **probability distribution**) if the following properties hold.
1. (Probability of truth) $\mathbb{P}(\mathsf{T}) = 1$.
2. (Countable additivity) If $s_1, s_2, \ldots$ are a disjoint sequence of events, then

$$\mathbb{P}\left(\bigvee_{i=1}^{\infty} s_i\right) = \sum_{i=1}^{\infty} \mathbb{P}(s_i).$$

FIVE MORE RULES OF PROBABILITY

So why make a formal definition of probability? Well, it turns out that many other rules that would be useful in calculating probabilities can be derived just from the two rules given above. Here are five in particular.

F8 For any probability function $\mathbb{P}$, and events $s_1, s_2, s_3, \ldots$, the following hold.
1. (False events) $\mathbb{P}(\mathsf{F}) = 0$.
2. (Disjoint finite events) If $s_1, \ldots, s_n$ are disjoint, then

$$\mathbb{P}\left(\bigvee_{i=1}^{n} s_i\right) = \sum_{i=1}^{n} \mathbb{P}(s_i).$$

3. (Negation rule) $\mathbb{P}(\neg s_1) = 1 - \mathbb{P}(s_1)$.
4. (Nondisjoint events) $\mathbb{P}(s_1 \vee s_2) = \mathbb{P}(s_1) + \mathbb{P}(s_2) - \mathbb{P}(s_1 s_2)$.

5. (Implication) If $s_1 \to s_2$, then $\mathbb{P}(s_1) \leq \mathbb{P}(s_2)$.

These five facts are also true for indicator functions $\mathbb{I}$! The fact that they also hold for probability functions is just icing on the cake.

To prove these facts about probability functions, it helps to do them in order.

> **WHY NOT JUST ASSUME FACTS?**
> At this point you might be wondering, why not just assume that $\mathbb{P}(\mathsf{F}) = 0$ instead of trying to prove it? There are two reasons. First, the goal is to assume as few things as possible, and then try to build out from there. Second, the more assumptions you make, the more likely you are to cause a contradiction in your assumptions, which brings the whole logical structure crashing down. So that is why it is standard operating procedure in mathematics to assume as little as possible, and see if we can get the things about probability that we wish were true as a logical conclusion from the assumptions.

PROBABILITY OF A FALSE STATEMENT IS 0

First show that $\mathbb{P}(\mathsf{F}) = 0$.

Proof. Let $\mathbb{P}(\mathsf{F}) = \alpha$. Then $s_1 = s_2 = \cdots = \mathsf{F}$ form a sequence of events satisfying

$$\sum_{i=1}^{\infty} \mathbb{I}(s_i) = \sum_{i=1}^{\infty} 0 = 0 \leq 1.$$

Hence the events are disjoint. By countable additivity,

$$\mathbb{P}\left(\bigvee_{i=1}^{\infty} s_i\right) = \sum_{i=1}^{\infty} \mathbb{P}(s_i)$$
$$= \mathbb{P}(s_1) + \sum_{i=2}^{\infty} \mathbb{P}(s_i)$$
$$= \mathbb{P}(s_1) + \mathbb{P}\left(\bigvee_{i=2}^{\infty} s_i\right).$$

But all the terms are the probability of a false statement, and so are just α, giving

$$\alpha = \alpha + \alpha$$

so $\alpha = 0$. □

FINITE DISJOINT EVENTS

Next comes the rule that for a finite set $s_1, \ldots, s_n$ of disjoint events,

$$\mathbb{P}\left(\bigvee s_i\right) = \sum \mathbb{P}(s_i).$$

Proof. Let $\{s_1, \ldots, s_n\}$ be any finite set of disjoint events, and then let $s_{n+1} = s_{n+2} = \cdots = \mathsf{F}.$. Then

$$\sum_{i=1}^{\infty} \mathbb{I}(s_i) = \sum_{i=1}^{n} \mathbb{I}(s_i) + \sum_{i=n+1}^{\infty} \mathbb{I}(s_i) \leq 1 + 0 = 1,$$

so the sequence is disjoint. So by countable additivity,

$$\mathbb{P}\left(\bigvee_{i=1}^{\infty} s_i\right) = \sum_{i=1}^{\infty} \mathbb{P}(s_i) = \sum_{i=1}^{n} \mathbb{P}(s_i) + \sum_{i=n+1}^{\infty} \mathbb{P}(s_i).$$

Since each $s_i = F$ for $i \geq n+1$, it holds that $(i \geq n+1) \to (\mathbb{P}(s_i) = 0)$, so

$$\sum_{i=n+1}^{\infty} \mathbb{P}(s_i) = 0.$$

Also,

$$\bigvee_{i=1}^{\infty} s_i = \left(\bigvee_{i=1}^{n} s_i\right) \vee \left(\bigvee_{i=n+1}^{\infty} s_i\right)$$
$$= \left(\bigvee_{i=1}^{n} s_i\right) \vee \mathsf{F} = \bigvee_{i=1}^{n} s_i$$

That means that

$$\mathbb{P}\left(\bigvee_{i=1}^{n} s_i\right) = \mathbb{P}\left(\bigvee_{i=1}^{\infty} s_i\right) = \sum_{i=1}^{n} \mathbb{P}(s_i),$$

and the proof is complete. □

CHAPTER 3: THE RULES OF PROBABILITY

Negation rule

Next is the negation rule, that

$$\mathbb{P}(\neg s) = 1 - \mathbb{P}(s).$$

Proof. Note that

$$\mathbb{I}(s) + \mathbb{I}(\neg s) = \mathbb{I}(s) + 1 - \mathbb{I}(s) = 1 = \mathbb{I}(\mathsf{T}).$$

Hence s and $\neg s$ are disjoint events and $(s \vee \neg s) = \mathsf{T}$.

Therefore, the previous rule gives

$$\mathbb{P}(s \vee \neg s) = \mathbb{P}(s) + \mathbb{P}(\neg s) = \mathbb{P}(\mathsf{T}) = 1,$$

and the proof follows from rearranging the terms. □

Inclusion-Exclusion

Earlier it was shown that for any two logical statements

$$\mathbb{I}(s \vee r) = \mathbb{I}(s) + \mathbb{I}(r) - \mathbb{I}(sr).$$

The same principle holds for events as well, and is called the *Principle of Inclusion-Exclusion*. To verify it, a logical diagram called a *Venn diagram* can help to visualize what is happening. This diagram makes us aware that $s \vee r$ is the disjoint logical OR of three parts, $s(\neg r)$, sr, and $(\neg s)r$. The left circle represents when s is true, while the right is when r is true. So $s \vee r$ is the area of the two circles combined.

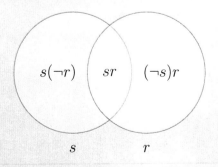

> **Logic and sets**
> Venn diagrams were initially introduced for logical statements, but today they are more commonly used to illustrate sets and set theory.

The Venn diagram informs the following fact.

> **F9** Given events s and r, then sr and $(\neg s)r$ are disjoint and $(sr \vee (\neg s)r) = r$.
> Furthermore, $s(\neg r)$, sr, and $(\neg s)r$ are disjoint and their logical OR is $s \vee r$.

Proof. Since $\mathbb{I}(s) + \mathbb{I}(\neg s) = 1$

$$\mathbb{I}(r) = \mathbb{I}(r)(\mathbb{I}(s) + \mathbb{I}(\neg s))$$
$$= \mathbb{I}(sr) + \mathbb{I}((\neg s)r).$$

The left hand side is at most 1 so the events in the right hand side terms must be disjoint. Further, because the sum of their indicators gives the indicator on the left, their logical OR must equal the event on the left.

To show the second part, recall

$$\mathbb{I}(s \vee r) = \mathbb{I}(s) + \mathbb{I}(r) - \mathbb{I}(s)\mathbb{I}(r)$$
$$= \mathbb{I}(s)(1 - \mathbb{I}(r)) + \mathbb{I}(r)$$
$$= \mathbb{I}(s)\mathbb{I}(\neg r) + \mathbb{I}(r)$$

□

When an event is the logical OR of disjoint events, it is called a partition.

> **D16** If $s = r_1 \vee r_2 \vee \cdots$ where the r_i are disjoint, say that the r_i **partition** s.

Now it is possible to show inclusion-exclusion for probabilities:

$$\mathbb{P}(s \vee r) = \mathbb{P}(s) + \mathbb{P}(r) - \mathbb{P}(sr).$$

Proof. Use the fact that $s \vee r$ can be partitioned into $s(\neg r)$, sr, and $(\neg s)r$ to write

$$\mathbb{P}(s \vee r) = \mathbb{P}((\neg s)r) + \mathbb{P}(sr) + \mathbb{P}(s(\neg r))$$

Adding and subtracting $\mathbb{P}(sr)$ gives

$$\mathbb{P}(s \vee r) = [\mathbb{P}((\neg s)r) + \mathbb{P}(sr)] + \\ [\mathbb{P}(s(\neg r)) + \mathbb{P}(sr)] - \mathbb{P}(sr).$$

Now use the fact that $(\neg s)r$ and sr partition r, and $s(\neg r)$ and sr partition s to get

$$\mathbb{P}(s \vee r) = \mathbb{P}(s) + \mathbb{P}(r) - \mathbb{P}(sr).$$

□

IMPLICATION

Recall that $s \to r$ means that whenever s is true, r must be true as well. If s is false, then r might be either true or false. So

$$(s \to r) = ((sr \vee \neg s))$$

Another way to say this is as follows.

> **F10** Implication gives the following:
>
> $$(s \to r) \to (sr = s).$$

Proof. Assume $(s \to r) = (sr \vee \neg s) = \mathsf{T}$. Break it into two cases.

Case 1: $s = \mathsf{T}$. Then $sr \vee \neg s = \mathsf{T}$ and $\neg s = \mathsf{F}$ so $sr = \mathsf{T}$. In this case, $sr = s$.

Case 2: $s = \mathsf{F}$. Then $sr = \mathsf{F}$, so $s = sr$ in this case as well.

Since these are the only two cases, the proof is complete. □

The diagram for implication is called an *Euler diagram*. Whenever s is true, r is true as well.

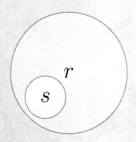

EULER AND VENN DIAGRAMS

It is easy to confuse *Euler* and *Venn* diagrams. Venn diagrams are always the most general situation. This means that any drawing with n logical statements will divide the paper into 2^n regions. For instance, s and r divide the logical space into $sr, (\neg s)r, s(\neg r),$ and $(\neg s)(\neg r)$. Euler diagrams can be about more specific relationships, as in the previous example of $s \to r$ where there are only three regions, $s, (\neg s)r,$ and $\neg r$.

If $s \to r$, then intuitively r is at least as likely to be true as s is, since whenever s is true so is r, but there could be cases where s is false but r is true. The implication rule makes this intuition precise,

$$(s \to r) \to (\mathbb{P}(s) \leq \mathbb{P}(r)).$$

Proof. As before,

$$\mathbb{P}(r) = \mathbb{P}(r(\neg s)) + \mathbb{P}(rs),$$

and since $s \to r$, $\mathbb{P}(rs) = \mathbb{P}(s)$.
So

$$\mathbb{P}(r) = \mathbb{P}(r(\neg s)) + \mathbb{P}(s),$$

and since $\mathbb{P}(r(\neg s)) \geq 0$, it holds that $\mathbb{P}(r) \geq \mathbb{P}(s)$. □

SOLVING THE STORY

In the story given at the beginning, let

$s = $ The group of monsters contains a Giant.
$r = $ The group of monsters contains an Ogre.

Assigning English phrases to one letter variable names allows us to write down the problem much more compactly. The information the party has is

$$\mathbb{P}(s) = 60\%$$
$$\mathbb{P}(r) = 70\%$$
$$\mathbb{P}(sr) = 40\%.$$

Therefore,

$$\mathbb{P}(s \vee r) = 0.6 + 0.7 - 0.4 = \boxed{90\%}.$$

EXAMPLES

E12 Find $\mathbb{P}((x < 3) \land (x > 4))$.
Answer. No matter the value of x, $(x < 3) \land (x > 4) = \mathsf{F}$, so $\mathbb{P}((x < 3) \land (x > 4)) = \boxed{0}$.

E13 If $\mathbb{P}(X = 3) = 0.2$, $\mathbb{P}(X = 4) = 0.3$, and $\mathbb{P}(X = 5) = 0.25$, what is $\mathbb{P}(X \in \{3, 4, 5\})$?
Answer. Since $(X = 3)$, $(X = 4)$, and $(X = 5)$ are disjoint events, $\mathbb{P}(X \in \{3, 4, 5\}) = \mathbb{P}(X = 3) + \mathbb{P}(X = 4) + \mathbb{P}(X = 5)$ so $\boxed{0.7500}$.

E14 If $\mathbb{P}(X \leq 4) = 0.1155$ where X is a real number, what is $\mathbb{P}(X > 4)$?
Answer. Since $(X \leq 4) = \neg(X > 4)$, $\mathbb{P}(X > 4) = 1 - \mathbb{P}(X \leq 4)$ and the answer is $\boxed{0.8845}$.

E15 Two friends go miniature golfing on Friday nights. The chance that the first player gets at least one hole in one is 30%. The chance that the second player gets at least one hole in one is 20%. The chance that both players get at least one hole in one is 10%. What is the chance that either one of the players (or both) gets a hole in one?
Answer. Let s_i denote the event that player i gets at least one hole in one. Then

$$\mathbb{P}(s_1 \lor s_2) = \mathbb{P}(s_1) + \mathbb{P}(s_2) - \mathbb{P}(s_1 s_2)$$
$$= 30\% + 20\% - 10\%,$$

and so the answer is $\boxed{40\%}$.

E16 If it is known that $\mathbb{P}(Y \geq 10) = 0.3$, what can you say about $\mathbb{P}(Y \geq 8)$?
Answer. Since $(Y \geq 10) \to (Y \geq 8)$, the best that can be said is that $\mathbb{P}(Y \geq 8)$ is $\boxed{\text{at least } 30\%}$.

E17 Suppose $X \sim \mathsf{d8}$. What is $\mathbb{P}(X > 6)$?
Answer. This can happen when $X = 7$ or $X = 8$. Since $X \sim \mathsf{d8}$ means the die is fair, each of these happen with probability $1/8$, so

$$\mathbb{P}(X > 7) = \mathbb{P}(X = 7) + \mathbb{P}(X = 8)$$
$$= 1/8 + 1/8 = 1/4$$

making the answer $\boxed{25\%}$.

ENCOUNTERS

33. TRUE STATEMENTS

For $X \in \mathbb{R}$, what is $\mathbb{P}(X^2 \geq 0)$?

34. MORE TRUE STATEMENTS

For X a real number, what is $\mathbb{P}((X > 2) \lor (X < 10))$?

35. FALSE STATEMENTS

For $Y \in \mathbb{R}$, what is $\mathbb{P}(|Y| < 0)$?

36. MORE FALSE STATEMENTS

For X a real number, what is $\mathbb{P}((X > 10) \land (X < 2))$?

37. ROLLING THE DICE

Consider the following.
a. If $W \sim \mathsf{d6}$, what is $\mathbb{P}(W \text{ is even})$?
b. If $R \sim \mathsf{d10}$, what is $\mathbb{P}(R \leq 7)$?
c. If $Y \sim \mathsf{d100}$, what is $\mathbb{P}(Y \leq 72)$?

38. A FAIR SIX-SIDED DIE

If $X \sim \mathsf{d6}$, what is
a. $\mathbb{P}(X = 1)$.
b. $\mathbb{P}(X \in \{1, 2\})$.
c. $\mathbb{P}(X \leq 3.5)$.
d. For $i \in \{1, 2, 3, 4, 5, 6\}$, $\mathbb{P}(X \leq i)$.

39. USING RULES

If $\mathbb{P}(Y < 2) = 0.3$ and $\mathbb{P}(Y \in [2, 3]) = 0.4$, what is $\mathbb{P}(Y \in (-\infty, 3])$?

40. Breaking apart problems

Suppose $\mathbb{P}(T = 1) = 0.2$, $\mathbb{P}(T = 2) = 0.15$, and $\mathbb{P}(T = 3) = 0.4$. What is $\mathbb{P}(T \in \{1, 2, 3\})$?

41. Negation rule

If $\mathbb{P}(G \geq 4) = 0.2$, what is $\mathbb{P}(G < 4)$?

42. More negation

If $\mathbb{P}(X = 4) = 0.3$, what is $\mathbb{P}(X \neq 4)$?

43. Implication

Given that $\mathbb{P}(R \leq 4) = 0.2$, what can be said about $\mathbb{P}(R \leq 5)$?

44. More implication

Suppose $\mathbb{P}(X = 3) = 0.6$. What can be said about $\mathbb{P}(X \in \{2, 3, 4\})$?

45. Combining rules

Suppose $\mathbb{P}(X \geq 3) = 0.4$, $\mathbb{P}(X \leq 4) = 0.8$. What is $\mathbb{P}(X \in [3, 4])$?

46. Inclusion-Exclusion

Suppose $\mathbb{P}(X \in A) = 0.4$, $\mathbb{P}(X \in B) = 0.3$, and $\mathbb{P}(X \in A \cap B) = 0.15$. What is $\mathbb{P}(X \in A \cup B)$?

Chapter 4: Conditional Probability

HE Ranger and the Thief looked at the small pile of gems and gold coins they had taken from the defeated Lich's lair. Both felt exhilarated, both felt exhausted. Each eyed the treasure, and then each other. "Feeling lucky?" the Thief said mischievously, and continued, "Then perhaps a game to decide the owner of the treasure? Winner take all!" The stakes were set. The game would be determined by three flips of a gold coin. They selected one golden coin from the hoard. The side with the Emperor Konravis imprinted upon it they dubbed heads, the other side with a wyvern they called tails. A head would count as a win for the Ranger, tails for the Thief. Out of three flips, the player who won two or more would take it all. The first flip was made, and heads came up! The Ranger smiled, knowing that the odds of winning had just increased. But by how much?

Just at that moment, the rest of the party burst in. "We must flee," they shouted, "Grab your riches and go!" The Ranger and Thief scooped their treasure into a pouch. But who owned it? The Ranger was certainly favored to win, having won the first game, but it was not impossible that the Thief could still win the game should it continue later.

The two players faced the *Problem of Points*. Given that the Ranger had taken the first point, but that the game had not concluded, how should they now fairly divide the loot to acknowledge that fact?

Partial Information

Consider the following event:

s = The Ranger wins the three coin flips.

Assuming the coin is fair, before any flips have been made, the Principle of Indifference would lead us to say that either player is equally likely to win the game. So $\mathbb{P}(s) = 1/2$.

But after the first coin flip, our knowledge has increased. Let us give that event a name.

h_1 = The first coin flip was heads.

It is no longer the case that we have no information: instead we know that statement h_1 is true. That changes things! To reflect that fact, we need better notation.

Given Information

The probability that statement s is true given that statement r is true will be denoted

$$\mathbb{P}(s \mid r).$$

This is read: "the probability s is true given that r is true."

Everything to the right of the vertical bar | is information we have, everything to the left of the vertical bar is things that are still uncertain. Important note: in general $\mathbb{P}(s \mid r) \neq \mathbb{P}(r \mid s)$. Never just flip the order in conditional probability!

So now there is notation for conditional probabilities, but how can they actually be calculated? Well, two of them are easy. Note that

$$\mathbb{P}(r \mid r) = 1$$
$$\mathbb{P}(\neg r \mid r) = 0.$$

The first equation says that conditioned on r being true, r is true and so has probability 1. The second line says that conditioned on r being true, $\neg r$ is false and so has probability 0. So far, so good.

But what about more general $\mathbb{P}(s \mid r)$ where event s is not completely determined by event r? It helps to consider what happens in the game played by the Ranger and the Thief, the Problem of Points.

Solving The Problem of Points

In a game of three flips where each coin has two outcomes, there are $2 \cdot 2 \cdot 2 = 8$ different outcomes that could occur. Recall s is the outcome that the Ranger wins the game, so there are at least two heads. The event h_1 is that the first outcome is a head.

Outcomes of the Problem of Points

Outcome	s	h_1	$s \wedge h_1$
(H,H,H)	T	T	T
(H,H,T)	T	T	T
(H,T,H)	T	T	T
(H,T,T)	F	T	F
(T,H,H)	T	F	F
(T,H,T)	F	F	F
(T,T,H)	F	F	F
(T,T,T)	F	F	F

If the coin is fair, then each of these outcomes is equally likely. But if the first flip is known to be a head, then only the first four outcomes in the table are possible. The rest can be ignored. And out of those first four outcomes, which are HHH, HHT, HTH, and HTT, three out of four result in the Ranger winning the overall game. Hence $\mathbb{P}(s \mid h_1) = 3/4$ in this case!

The probability the Ranger wins overall will be $\boxed{75\%}$, which would indicate that unless they are able to complete the game, the Ranger should get 75% of the treasure they were betting on.

The General Formula

To generalize this notion, consider a figure where the probability of an event is proportional to its size. There are two events to consider, s and r.

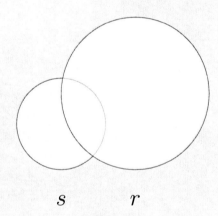

Given that r is known to be true, probabilities are now thought of with the notion that $\mathbb{P}(r \mid r) = 1$. Hence either the event sr or the event $(\neg s)r$ must be true, there are no other options. Then the proportion of area out of r given to sr is

$$\mathbb{P}(sr \mid r) = \frac{\mathbb{P}(sr)}{\mathbb{P}(sr) + \mathbb{P}(\neg sr)} = \frac{\mathbb{P}(sr)}{\mathbb{P}(r)}$$

In general, this idea is called the *conditional probability formula*.

D17 The conditional probability formula

says that for $\mathbb{P}(r) > 0$,

$$\mathbb{P}(s \mid r) = \frac{\mathbb{P}(sr)}{\mathbb{P}(r)}.$$

For the Problem of points, $\mathbb{P}(sr) = 3/8$ (since 3 out of the 8 cases have both the Ranger winning the first flip and winning overall) and $\mathbb{P}(r) = 4/8$ (since 4 out of the 8 cases have the Ranger winning the first flip). Hence

$$\mathbb{P}(s \mid r) = \frac{sr}{r} = \frac{3/8}{4/8} = \frac{3}{4}$$

as before.

E18 The chance that a student both studies for and does well on an exam is 40%. The chance that the student does well on the exam is 50%. What is the chance that the student studies given that they did well on the exam?
Answer. Let s be the event that the student studies and w be the event that the student does well on the exam. Then

$$\mathbb{P}(w) = 0.5$$

and

$$\mathbb{P}(sw) = 0.4,$$

so

$$\mathbb{P}(s \mid w) = \frac{\mathbb{P}(sw)}{\mathbb{P}(w)} = \frac{0.4}{0.5},$$

which is $\boxed{80\%}$.

E19 A biologist is studying two traits in a population. If 90% of the species has the first trait and 85% has both traits, what is the probability that a subject with the first trait has both traits?
Answer. In these types of problems, it is important to write down the information that is given. First, name the events of importance.

$$t_1 = \text{has the first trait}$$
$$t_2 = \text{has the second trait}$$

Next, what information does the problem give?

$$\mathbb{P}(t_1) = 90\%$$
$$\mathbb{P}(t_1 t_2) = 85\%.$$

Finally, what is the problem asking for? In this case,

$$\mathbb{P}(t_2 \mid t_1).$$

Fortunately, the conditional probability formula only requires information given by the problem statement!

$$\mathbb{P}(t_2 \mid t_1) = \frac{\mathbb{P}(t_1 t_2)}{\mathbb{P}(t_1)} = \frac{85}{90},$$

which is approximately $\boxed{94.44\%}$.

> **CONDITIONAL KEYWORDS**
> Keywords for conditional probability are "given", "when", "assuming", and "if-then". For instance, "What is the probability of snow given that it snowed yesterday?", "What is the probability of snow when it snowed yesterday?", "What is the probability of snow today assuming it snowed yesterday", and "If it snowed yesterday, then what is the probability of snow today" are all asking "What is $\mathbb{P}(s_1 \mid s_0)$?" where s_1 is the event that it snows today, and s_0 is the event that it snows yesterday.

Remember, when solving probability problems, the following steps can be helpful.

> **STEPS FOR SOLVING PROBABILITY PROBLEMS**
> **First.** Write down what you know. This means give names to events that are part of the problem, and write down the probabilities and conditional probabilities that you are given.
> **Second.** Write down the goal using mathematics, that is, in probability notation.
> **Third.** Only after completing steps 1 and 2 are you ready to tackle the problem!

TWO-STAGE EXPERIMENTS

Another way to think of the conditional probability formula is by viewing the problem as a *two-stage experiment*. Think about the event that both s and

r are true. Then first r has to hold, which happens $\mathbb{P}(r)$ fraction of the time. Out of the times that r holds, the fraction of the time that s holds will be $\mathbb{P}(s \mid r)$. So $\mathbb{P}(r)\mathbb{P}(s \mid r)$ should be the probability that s and r hold, or $\mathbb{P}(sr)$. That is,

$$\mathbb{P}(r)\mathbb{P}(s \mid r) = \mathbb{P}(sr).$$

That is just a rearrangement of the conditional probability formula.

> **E20** The chance of rain tomorrow is 80%. The chance of lightning given that it rains tomorrow is 10%. What is the chance of both rain and lightning?
> **Answer.** If a is the event that it rains, and b the event that there is lightning, then
>
> $$\mathbb{P}(ab) = \mathbb{P}(a)\mathbb{P}(b \mid a) = (80\%)(10\%)$$
>
> which is 8%.

There is no particular reason to stop at two stages of an experiment. This idea works for three or more stages as well:

$$\mathbb{P}(a_1 a_2 a_3) = \mathbb{P}(a_1)\mathbb{P}(a_2 \mid a_1)\mathbb{P}(a_3 \mid a_2 a_1).$$

In this fashion, the probability that any number of events that are all true can be calculated!

ODDS

Another way to describe probabilities is through the use of *odds*.

> **D18** For a given statement s, the **odds** that s is true is
>
> $$\frac{\mathbb{P}(s)}{\mathbb{P}(\neg s)}.$$

For example, if $\mathbb{P}(s) = 2/3$, then $\mathbb{P}(\neg s) = 1/3$, and the odds that s happens is $(2/3)/(1/3) = 2/1$. This is often read as "2 to 1" odds, or using an older notation for fractions, $2:1$.

Unlike probabilities, odds can be greater than 1 (for instance, $5:4$), less than 1 ($3:7$), or equal 1 ($1:1$ odds).

ENCOUNTERS

47. A CONDITIONAL DIE

Suppose $X \sim d6$. Find $\mathbb{P}(X = 1 \mid X \leq 4)$.

48. BASIC CONDITIONAL PROBABILITY

Suppose $\mathbb{P}(p) = 0.3$, $\mathbb{P}(q) = 0.2$ and $\mathbb{P}(p, q) = 0.15$.
a. What is $\mathbb{P}(p \mid q)$?
b. What is $\mathbb{P}(q \mid p)$?

49. INTERVAL CONDITIONING

The probability that $X \in [3, 7]$ is 0.423 and the probability that $X \leq 7$ is 0.620. What is $\mathbb{P}(X \in [3, 7] \mid X \leq 7)$?

50. ANOTHER CONDITIONAL DIE

For $Y \sim d10$, what is $\mathbb{P}(Y = 4 \mid Y \leq 8)$?

51. BLOOD TESTS

Given that a patient has a particular disease, the chance that a particular blood test comes back positive is 75%. The chance that the blood test comes back positive if the patient does not have the disease is 10%. The chance of having the disease is 3%. What is the chance that a patient gets back a positive?

52. LOOKING FOR COPPER

A mining company models that if a hillside contains ore, there is a 20% chance that they will find it. If there is no ore, then of course they will not find any. Suppose there is a 36% chance the hillside contains ore. What is the chance that the company finds it?

53. The satellite

A weather satellite detects precipitation when precipitation exists 98% of the time. It gives a false positive and reports precipitation when none exists 4% of the time. If there is a 10% chance of precipitation, what is the chance that the satellite reports that there is precipitation?

54. Emergency!

A disaster preparation center classifies disasters as Type I, II, or III. The probability of each occurring in a decade is 5%, 2%, and 1% respectively.

Given that a Type I, II, or III disaster did occur, what is the probability that it is a Type I?

55. Raining once more

If the odds that it will rain today are 4 to 5, what is the probability that it will rain today?

56. Tumor odds

The odds of a malignant tumor metastasizing is 4:3. What is the probability that the tumor metastasizes?

57. Odds and fairness

Prove that if the odds for s are greater than 1, then $\mathbb{P}(s) > 50\%$.

58. Odds and probabilities

Prove that if the odds for s are greater than 4 to 1, then $\mathbb{P}(s) > 80\%$.

59. Experience

Two out of seven staff members have experience working with R. If two staff members are chosen uniformly to be part of a task force, what is the chance that both will know R?

60. Drawing cards

In a standard deck of 52 cards, 13 are hearts.
a. If two cards are chosen from the deck without replacement, what is the chance that both are hearts?
b. What is the chance that neither card is a heart?
c. What is the chance that exactly one card is a heart?

61. The food pantry

There are 9 cans in a food pantry that are unlabeled, but the inventory sheet says that they must be 4 cans of green beans and 5 cans of corn.
a. If 4 cans are chosen uniformly without replacement, and the first two cans chosen have green beans, what is that chance that the third can has green beans?
b. If 4 cans are chosen uniformly without replacement, what are the chances that all 4 have green beans?

62. A corny problem

Continuing the last problem, what is the chance that all 4 of the cans chosen have corn?

Chapter 5: Independence

BOUNTY HUNTERS WERE following the Pirate Captain's ship. There were five different bounty hunters, each working independently. If each bounty hunter had only a 20% chance of catching the Captain's ship, then what was the chance that at least one of the bounty hunters caught up?

Independence of Two Events

Sometimes, information about whether or not an event like q is true does not give us any information about whether or not event p is true. In this case, call the two events *independent*. Note that if $\mathbb{P}(p \mid q) = \mathbb{P}(p)$, then $\mathbb{P}(p) = \mathbb{P}(pq)/\mathbb{P}(q)$ and $\mathbb{P}(p)\mathbb{P}(q) = \mathbb{P}(pq)$. That makes for a better definition of independence, since it works even when $\mathbb{P}(p)$ or $\mathbb{P}(q)$ is 0.

D19 Events p and q are **independent** if
$$\mathbb{P}(pq) = \mathbb{P}(p)\mathbb{P}(q).$$

This is a perfectly symmetric definition.

F11 If p and q are independent, then so are q and p.

Proof. Say p and q are independent, then
$$\mathbb{P}(qp) = \mathbb{P}(pq) = \mathbb{P}(p)\mathbb{P}(q) = \mathbb{P}(q)\mathbb{P}(p).$$
□

Another way to talk about independence is in terms of conditional probability. Event p is independent of q if conditioning on q being true does not change the probability that p is true.

F12 Suppose p and q are events with $\mathbb{P}(q) > 0$. Then p and q are independent if and only if $\mathbb{P}(p \mid q) = \mathbb{P}(p)$.

If p and q are independent, it means that knowledge of one has no effect on the other. Therefore, it stands to reason that p and $\neg q$ should also be independent because knowing if q is not true should also not affect p.

F13 If p and q are independent, then so are p and $\neg q$.

Proof. Suppose p and q are independent. Then $pq \vee p(\neg q)$ form a disjoint partition of p, so
$$\mathbb{P}(p) = \mathbb{P}(pq \vee p(\neg q)) = \mathbb{P}(pq) + \mathbb{P}(p(\neg q)).$$

Rearranging gives
$$\begin{aligned}\mathbb{P}(p(\neg q)) &= \mathbb{P}(p) - \mathbb{P}(pq) \\ &= \mathbb{P}(p) - \mathbb{P}(p)\mathbb{P}(q) \\ &= \mathbb{P}(p)(1 - \mathbb{P}(q)) \\ &= \mathbb{P}(p)\mathbb{P}(\neg q),\end{aligned}$$

so p and $\neg q$ are also independent. □

More than two events

So what should it mean for events $s_1, s_2, \ldots, s_n$ to be independent? What should independence of a sequence $s_1, s_2, \ldots$ mean? The idea is to consider all finite subsets of the set of events. The probability that every one of the finite subset of events holds equals the product of the probabilities that each event holds.

> **D20** A finite set of events $\{s_1, \ldots, s_n\}$ is independent if for all $I \subseteq \{1, \ldots, n\}$,
> $$\mathbb{P}\left(\bigwedge_{i \in I} s_i\right) = \prod_{i \in I} \mathbb{P}(s_i).$$

This always holds when I has no events or just one event, but it must also hold when I contains two or more events. So the number of requirements grows quickly in n. For $n = 3$ and events s_1, s_2, s_3, this requires that

$$\mathbb{P}(s_1 s_2) = \mathbb{P}(s_1)\mathbb{P}(s_2)$$
$$\mathbb{P}(s_1 s_3) = \mathbb{P}(s_1)\mathbb{P}(s_3)$$
$$\mathbb{P}(s_2 s_3) = \mathbb{P}(s_2)\mathbb{P}(s_3)$$
$$\mathbb{P}(s_1 s_2 s_3) = \mathbb{P}(s_1)\mathbb{P}(s_2)\mathbb{P}(s_3).$$

Note that it is possible for the last equation to be satisfied, but not have one of the first three hold true. When $n = 4$, there are 11 such equations, and when $n = 5$, there are 26. So usually this works in the reverse direction: rather than using the equations to prove independence, independence is assumed and then the equations can be used to break any logical AND into multiplied probabilities.

> **Disjoint versus independent**
> Disjoint events allow probabilities of logical OR to be turned into sums. Independent events allow probabilities of logical AND to be turned into products. They should not be confused! In particular, disjoint events with probabilities strictly between 0 and 1 mean that at most one of the events can occur, so they are never independent.

A similar result as before holds for negations.

> **F14** If $s_1, \ldots, s_n$ are independent, so are $f_1, \ldots, f_n$, where for each f_i either $f_i = s_i$ or $f_i = \neg s_i$.

Proof. Use induction on m, the number of i such that $f_i = \neg s_i$. The base case when $m = 0$ is trivial since the s_i and the f_i are the same.

Now suppose that the statement holds for some m and consider a new set of f_i with $m + 1$ negations.

Let $k = \max\{i : f_i = \neg s_i\}$. Set $g_i = f_i$ for $i \neq k$ and $g_k = s_k$.

Then by the two events result, s_k and
$$\bigwedge_{i \neq k} f_i$$
are independent, which means $f_k = \neg s_k$ and $\bigwedge_{i \neq k} f_i$ are independent so

$$\mathbb{P}\left(\bigwedge f_i\right) = \mathbb{P}\left(f_k \bigwedge_{i \neq k} f_k\right) = \mathbb{P}(f_k)\mathbb{P}\left(\bigwedge_{i \neq k} f_k\right).$$

By the induction hypothesis,
$$\mathbb{P}\left(\bigwedge_{i \neq k} f_i\right) = \prod_{i \neq k} \mathbb{P}(f_i),$$
which combined with the $\mathbb{P}(f_k)$ factor completes the induction. □

Sequences

So how should the independence of an infinite sequence of events be handled? This is done by requiring that any finite subset of the sequence of events should be independent.

D21 A sequence $s_1, s_2, \ldots$ of events is **independent** if for every $n \in \{1, 2, 3, \ldots\}$, the events $\{s_1, s_2, \ldots, s_n\}$ are independent.

It is actually pretty rare to have to prove directly that a sequence is independent. Instead, it is often something assumed based on the situation, and then the property can be used.

Solving the Story

Back to the pirate and bounty hunters! Let s_i denote the event that the ith bounty hunter catches the Pirate Captain. Then the Captain is interested in the probability that at least one of the bounty hunters catches the ship. That is, the goal is to find

$$\mathbb{P}\left(\bigvee_{i=1}^{5} s_i\right).$$

The problem implies that the s_i are independent, but independence only applies to a logical AND. The problem is stated using logical OR. The solution is to use De Morgan's laws to switch from OR to AND. De Morgan's laws state that

$$\neg\left(\bigvee_{i=1}^{5} s_i\right) = \bigwedge_{i=1}^{5} \neg s_i,$$

so

$$\left(\bigvee_{i=1}^{5} s_i\right) = \neg\bigwedge_{i=1}^{5} \neg s_i,$$

Applied to our Captain's question:

$$\mathbb{P}\left(\bigvee_{i=1}^{5} s_i\right) = \mathbb{P}\left(\left(\neg \bigwedge_{i=1}^{5} \neg s_i\right)\right)$$

$$= 1 - \left(\prod_{i=1}^{5} \mathbb{P}(\neg s_i)\right)$$

$$= 1 - (1 - 0.2)^5$$

$$= 0.67232.$$

So our bounty hunters have about a 67.23% chance that at least one of them catches the Captain's ship.

Note on Independence of Three or More Events

It is tempting to think that

$$\mathbb{P}(s_1 s_2 \cdots s_n) = \prod_{i=1}^{n} \mathbb{P}(s_i),$$

is enough to give independence of s_1, s_2, and s_3, but it is not. It really is necessary to have the more complicated definition.

To see why, suppose $X \sim \text{d}27$ is a roll of a fair 27 sided die. Suppose $s_1 = (X \leq 18)$, $s_2 = (X \geq 10)$, and

$$s_3 = (x \geq 11) \vee (X = 1).$$

Then

$$\mathbb{P}(s_1) = \mathbb{P}(s_2) = \mathbb{P}(s_3) = \frac{18}{27} = \frac{2}{3},$$

and $s_1 s_2 s_3 = (X \in \{11, \ldots, 18\})$ so

$$\mathbb{P}(s_1, s_2, s_3) = \frac{8}{27} = \left(\frac{2}{3}\right)^3 = \mathbb{P}(s_1)\mathbb{P}(s_2)\mathbb{P}(s_3).$$

However, s_1, s_2, and s_3 are not independent! There are many reasons why, one is that s_1 and s_2 are not independent.

$$\mathbb{P}(s_1 s_2) = \frac{9}{27},$$

while

$$\mathbb{P}(s_1)\mathbb{P}(s_2) = \frac{2}{3}\frac{2}{3} = \frac{4}{9} = \frac{12}{27}.$$

So, just because the logical AND of three events breaks into the product of the probabilities does not mean that the logical AND of any two of the events will break into smaller products.

Encounters

63. Independence of two events

Suppose that $\mathbb{P}(a_1) = 0.3$ and $\mathbb{P}(a_2) = 0.6$, where a_1 and a_2 are independent events. What is $\mathbb{P}(a_1 a_2)$?

64. Independence of three events

Suppose that $\mathbb{P}(r) = \mathbb{P}(s) = 0.4$ and $\mathbb{P}(t) = 0.6$ where r, s, and t are independent events. What is $\mathbb{P}(rst)$?

65. All false

Suppose that s_1, s_2, s_3 are independent events each with probability 0.2.
a. What is the chance that none of the events occur?
b. What is the chance that at least one of the events occurs?

66. Gold!

Four mining companies are searching for gold independently in a range of mountains. If each has a 10% of finding gold, what is the chance that at least one of them does so?

67. The archers

Four archers independently fire at a target. Each has a 0.2 chance of striking the target.
a. What is the chance that all the archers miss the target?
b. What is the chance that at least one archer hits the target?

68. Finding a vaccine

Five labs are independently working on a vaccine for a particular disease. If each lab has a 60% chance of developing a working vaccine, what is the chance that at least one lab succeeds?

69. Factory woes

On a given day in a factory, there is a 3% chance of a shutdown, a 1% chance of a worker injury, and a 2% chance of a delivery delay. If these three events are independent, what is the chance that all three occur on a given day?

70. High school success

At a particular high school, each senior student has a 88% chance of graduating. If there are 100 students in the senior class, what is the chance that everyone graduates?

Chapter 6: Bayes' Rule

The Scout waited for one of two monster armies to approach. The first monster army consisted of 60% orcs, 30% ogres, and 10% giants. The second monster army consisted of 90% orcs and 10% ogres. Initially, the Scout believed that the army that was approaching was equally likely to be one of the two armies. Then the Scout saw an orc emerge from the woods. Now what was the chance that the approaching army was the first monster army?

Flipping Conditioning

A pretty common mistake when working with conditioning is to assume that $\mathbb{P}(s \mid r)$ is the same as $\mathbb{P}(r \mid s)$.

Here is a simple example to see why this fails. Suppose $X \sim d6$ is a roll of a fair six sided die. Then let s be the event that $X \geq 2$, and r be the event that $X \leq 4$. Then the probability of s given information r is

$$\begin{aligned}\mathbb{P}(s \mid r) &= \mathbb{P}(X \geq 2 \mid X \leq 4) \\ &= \frac{\mathbb{P}(X \geq 2, X \leq 4)}{\mathbb{P}(X \leq 4)} \\ &= \frac{3/6}{4/6} = \frac{3}{4} = 75\%.\end{aligned}$$

On the other hand, the probability of r given information s is

$$\begin{aligned}\mathbb{P}(r \mid s) &= \mathbb{P}(X \leq 4 \mid X \geq 2) \\ &= \frac{\mathbb{P}(X \geq 2, X \leq 4)}{\mathbb{P}(X \geq 2)} \\ &= \frac{3/6}{5/6} = \frac{3}{5} = 60\%.\end{aligned}$$

The *numerator* will be the same for $\mathbb{P}(s \mid r)$ and $\mathbb{P}(r \mid s)$, but the *denominator* will be different for the two problems. Of course, to convert from one to the other, it is just necessary to multiply and divide by the appropriate denominators. This technique is known as *Bayes' Rule*.

> **T2** Bayes Rule
> If s and r are events with positive probability,
> $$\mathbb{P}(s \mid r) = \mathbb{P}(r \mid s)\frac{\mathbb{P}(s)}{\mathbb{P}(r)}.$$

> **Bayes' Rule**
> Bayes' Rule is named after Thomas Bayes, an 18th century statistician and Presbyterian minister who wanted to understand how obtaining evidence affected probabilities of events.

For such an important theorem, the proof is very easy.

Proof. Let s and r be two events of nonzero probability. Then

$$\mathbb{P}(sr) = \mathbb{P}(s \mid r)\mathbb{P}(r) = \mathbb{P}(r \mid s)\mathbb{P}(s).$$

Dividing by $\mathbb{P}(r)$ finishes the proof. □

Solving the Story

Let s denote the event that the approaching army is the first monster army, and r the event that the first monster seen in the army is an Orc. With this information, our goal is to find
$$\mathbb{P}(s|r)$$
The problem reveals three pieces of information:
$$\mathbb{P}(s) = 1/2,$$
$$\mathbb{P}(r \mid s) = 0.6,$$
$$\mathbb{P}(r \mid \neg s) = 0.9.$$

Therefore, it is much more likely to have seen an Orc if it was not the first monster army that approached. So it would seem that having this piece of information r should decrease the probability of s. Using the notation, this means that
$$\mathbb{P}(s|r) < \mathbb{P}(s).$$

To see if this is true, Bayes' Rule can be used to find $\mathbb{P}(s \mid r)$ from $\mathbb{P}(r \mid s)$. To use Bayes' Rule, it is necessary to know both $\mathbb{P}(s)$ and $\mathbb{P}(r)$. The value of $\mathbb{P}(s)$ is given in the problem statement, however, one piece of the puzzle is missing: what is $\mathbb{P}(r)$?

To get this, first divide the event r into the (disjoint) situation where both r and s occur, and the situation where r occurs but s does not:
$$\mathbb{P}(r) = \mathbb{P}(rs) + \mathbb{P}(r(\neg s)).$$

Now use the conditional probability formula to write these in terms of probabilities given in the Story.
$$\mathbb{P}(r) = \mathbb{P}(r \mid s)\mathbb{P}(s) + \mathbb{P}(r \mid \neg s)\mathbb{P}(\neg s).$$

Now put in numbers to get:
$$\mathbb{P}(r) = (0.6)(1/2) + (0.9)(1/2) = 0.75.$$

Finally, Bayes' Rule can be used!
$$\mathbb{P}(s \mid r) = \mathbb{P}(r \mid s)\frac{\mathbb{P}(s)}{\mathbb{P}(r)} = 0.6\frac{1/2}{3/4} = 0.4,$$

or $\boxed{40\%}$.

Using Bayes' Rule

To tackle Bayes' rule type of problems (or probability problems in general,) the following steps can be helpful.

a. Give names to all the events given in the problem.

b. Write down for these events all the probabilities given by the problem.

c. Write down the goal of the problem.

d. Use the probability formulas and Bayes' Rule to move from the information given to the goal.

The following examples illustrate this approach. The first example considers the problem of a medical trial of a drug.

E21 Suppose that a drug is believed to be either effective 20% of the time (probability 20%) or 50% of the time (probability 80%). The drug is administered independently to five trial patients, but the drug is ineffective for all trial patients. Given this information, what is the chance that the drug is effective 20% of the time?

Answer. Let s_{20} be the event that the drug is effective 20% of the time. Let f be the event that the drug is ineffective five times in a row. Then the goal is to find
$$\mathbb{P}(s_{20} \mid f).$$

First consider what is known. $\mathbb{P}(s_{20}) = 0.2$ with no information. If the drug is effective 20% of the time, the chance of failure five times in a row is 0.8^5. If the drug is effective 50% of the time, the chance of failure five times in a row is 0.5^5. Hence
$$\mathbb{P}(f \mid s_{20}) = 0.8^5$$
$$\mathbb{P}(f \mid \neg s_{20}) = 0.5^5.$$

To use Bayes' Rule to find $\mathbb{P}(s_{20} \mid f)$, it is also necessary to know what $\mathbb{P}(f)$ is. As before, break down the event into pieces to find this:

$$\mathbb{P}(f) = \mathbb{P}(fs_{20}) + \mathbb{P}(f(\neg s_{20}))$$
$$= \mathbb{P}(s_{20})\mathbb{P}(f \mid s_{20}) + \mathbb{P}(\neg s_{20})\mathbb{P}(f \mid \neg s_{20})$$
$$= (0.2)(0.8)^5 + (0.8)(0.5)^5$$
$$= 0.090536.$$

Hence Bayes' Rule gives

$$\mathbb{P}(s_{20} \mid f) = \mathbb{P}(f \mid s_{20})\frac{\mathbb{P}(s_{20})}{\mathbb{P}(f)}$$
$$= (0.2)(0.8)^5/0.090536,$$

which is about $\boxed{0.7238}$.

The next example also comes from the medical arena, but is focused on tests for disease.

E22 A particular disease occurs in about 1 in every 100,000 people in a population. A test for the disease returns true if you have the disease with probability 90%. The test makes a mistake and returns true even if you do not have the disease with probability about 1%. Given that the test returns positive, what is the chance that you actually have the disease?
Answer. Let d be the event that you have the disease, and t be the event that the test returns a positive result. Then the goal is to find $\mathbb{P}(d \mid t)$. The information given in the problem is

$$\mathbb{P}(d) = 10^{-5}$$
$$\mathbb{P}(t \mid d) = 0.9$$
$$\mathbb{P}(t \mid \neg d) = 0.01$$

To use Bayes' Rule, it is necessary to also find $\mathbb{P}(t)$. For this, once more use the fact that td and $t(\neg d)$ partition t. In words, t occurs when either both t and d occur or t and $\neg d$ occur, and td and $t(\neg d)$ cannot both occur at the same time. Hence

$$\mathbb{P}(t) = \mathbb{P}(td) + \mathbb{P}(t(\neg d))$$
$$= \mathbb{P}(t \mid d)\mathbb{P}(d) + \mathbb{P}(t \mid \neg d)\mathbb{P}(\neg d)$$
$$= (0.9)10^{-5} + (0.01)(1 - 10^{-5}) = 0.0100089$$

Then Bayes' Rule can be used:

$$\mathbb{P}(d|t) = \mathbb{P}(t|d)\frac{\mathbb{P}(d)}{\mathbb{P}(t)}$$
$$= 0.9\frac{10^{-5}}{0.0100089},$$

which is $\boxed{0.0008991}$ to four significant digits.

Because the test has a 90 times higher chance of returning true when you have the disease than not, the result is about 90 times the original chance of having the disease. But since the original chance of having the disease is so small, this is still a much greater chance that the test simply erred rather than actually having the disease.

Remember that there are two different ways to get a positive result: you have the disease and the test is correct, or you do not have the disease and the test is incorrect. Since the chance of actually having the disease is 10^{-5} and the chance that the test is incorrect is much higher (10^{-2}), Bayes' Rule shows that the most likely reason the test would show that you had the disease is that the test made a mistake, not that you actually have the disease (because it is so rare). That is the reason for the medical phrase "we need to run more tests". A first positive test is usually not solid proof that the subject is positive, but indicates that more accurate tests are probably a good idea to rule it out.

TERMINOLOGY

When calculating $\mathbb{P}(s \mid r)$, the event r occurring can be viewed as *evidence* in favor (or disfavor) of s. Prior to learning that r occurred, the probability of s is just $\mathbb{P}(s)$. So this is called the *prior* probability. After learning that r occurred, the probability is $\mathbb{P}(s \mid r)$. So this is the *posterior* probability.

> **D22** In considering $\mathbb{P}(s \mid r)$, call $\mathbb{P}(s)$ the prior probability of s, and $\mathbb{P}(s \mid r)$ the posterior probability of s given r.

THE LAW OF TOTAL PROBABILITY

This idea of finding the probability of an event by breaking it up into multiple events is sometimes called the *law of total probability*.

> **F15** The Law of Total Probability
> Suppose $s_1, s_2, \ldots$ are a sequence of disjoint events such that $\mathbb{P}\left(\vee_{i=1}^{\infty} s_i\right) = 1$, then
> $$\mathbb{P}(r) = \sum_{i=1}^{\infty} \mathbb{P}(r \mid s_i)\mathbb{P}(s_i).$$

Before proving this, it helps to have the following result.

> **F16**
> For any event a, if $\mathbb{P}(b) = 1$, then
> $$\mathbb{P}(a) = \mathbb{P}(ab).$$

Proof. Note that $\mathbb{P}(b) = 1$ means $\mathbb{P}(\neg b) = 1 - 1 = 0$, and that $b \vee \neg b = \mathsf{T}$, so
$$r = r(b \vee \neg b) = rb \vee r(\neg b).$$
So
$$\mathbb{P}(r) \leq \mathbb{P}(rb) + \mathbb{P}(r(\neg b)) = \mathbb{P}(rb).$$
But
$$rb \to r,$$
so
$$\mathbb{P}(r) \geq \mathbb{P}(rb),$$
and combining gives $\mathbb{P}(r) = \mathbb{P}(rb)$. □

Now the Law of Total Probability can be shown.

Proof of the Law of Total Probability. By the last result,
$$\begin{aligned}\mathbb{P}(r) &= \mathbb{P}(r\,(\vee_{i=1}^{\infty} s_i)) \\ &= \mathbb{P}(\vee_{i=1}^{\infty} r s_i) \\ &= \sum_{i=1}^{\infty} \mathbb{P}(r s_i) \\ &= \sum_{i=1}^{\infty} \mathbb{P}(r \mid s_i)\mathbb{P}(s_i)\end{aligned}$$

□

E23 Suppose for all $i \in \{1, 2, \ldots\}$ it holds that $\mathbb{P}(X = i) = (1/2)^i$. Suppose $\mathbb{P}(r \mid X = i) = (2/3)^i$. what is $\mathbb{P}(r)$?

Answer. The events $(X = i)$ are disjoint. Since $\sum_{i=1}^{\infty} (1/2)^i = 1$, the law of total probability applies.
$$\begin{aligned}\mathbb{P}(r) &= \sum_{i=1}^{\infty} \mathbb{P}(r \mid X = i)\mathbb{P}(X = i) \\ &= \sum_{i=1}^{\infty} (2/3)^i (1/2)^i \\ &= \sum_{i=1}^{\infty} (1/3)^i \\ &= \frac{1/3}{1 - 1/3} = \frac{1}{2},\end{aligned}$$
so the event r occurs with probability $\boxed{50\%}$.

The Law of Total Probability was written for an infinite sequence of events, but by using the trick of making $s_{n+1} = s_{n+2} = \cdots = \mathsf{F}$, the law also applies to any finite set of events as well.

ENCOUNTERS

71. BAYES' RULE

Suppose that $\mathbb{P}(X = 1) = 0.5$, $\mathbb{P}(X = 2) = 0.3$, $\mathbb{P}(X = 3) = 0.2$, and that $\mathbb{P}(r \mid X) = 1/X$.

a. What is $\mathbb{P}(r)$?
b. What is $\mathbb{P}(X = 1 \mid r)$?

72. Probability vectors

Suppose $\mathbb{P}(Y = i) = v_i$ where v is the probability vector $(0.2, 0.4, 0.4)$. Next, suppose $\mathbb{P}(s \mid Y) = Y^2/9$. Then what is $\mathbb{P}(Y = 1 \mid s)$?

73. Machine problems

Machine A has a 1% chance of making a widget with an error, while Machine B has a 5% chance of error.

If Machine A makes 70% of the widgets and Machine B makes 30%, what is the chance that a widget with an error came from Machine A?

74. Building cars

An automotive company sources their parts from five companies. Each part is equally likely to have come from the each company. If the chance of the part being malformed is $0.5\%, 0.5\%, 1\%, 1\%, 2\%$ for each of the five companies, what is the chance that a malformed part came from the first company?

75. Cholesterol

Suppose that in a population, there is a 3% chance of having a genetic marker that doubles the chance of having high cholesterol. So if H is the event that the person has high cholesterol, and G is the event that they have the genetic marker, then

$$\mathbb{P}(H \mid G) = 2\mathbb{P}(H \mid G^C).$$

Given that someone has high cholesterol, what is the chance that they have the marker?

76. Genetics

Suppose that in a population, there is a 5% chance of having a genetic marker that multiplies the chance of having asthma by 1.1. So if A is the event that the person has asthma, and G is the event that they have the genetic marker, then

$$\mathbb{P}(A \mid G) = 1.1\mathbb{P}(A \mid G^C).$$

Given that someone has asthma, what is the chance that they have the marker?

77. A bit of a screw-up

A company obtains screws from three manufacturers, code names X, Y, and Z. About 80% of screws come from company X, 10% from company Y, and 10% from company Z. Company X has about 2% of their screws defective, while company Y has only 1% defective, and company Z has 3% defective. Given that a particular screw is defective, find the probability that it came from company X, company Y, and company Z.

78. Getting in a bind

Antony, Boethius, and Cato are bookbinders working for a publishing company. Each produces about the same number of books. Antony has a 75% chance of producing a high quality book, Boethius 40%, and Cato 60%. Given that a book is of high quality, what is that chance that it was bound by Antony?

Chapter 7: Random Variables

HE RULER OF DAVORNIA SENT five emissaries to neighboring lands to seek out help against the pirates of the Birynian Sea. The Ruler believed that each emissary had an 80% chance independent of the others of making it to their destination successfully. What is the distribution of the number of emissaries that successfully reach their destination?

Random Variables

Our focus up until now has been on *events*; now consider mathematical objects that are called *random variables*.

D23 Say that X is a **random variable** if there exists a collection of sets $\mathcal{F}_X$ such that $\mathbb{P}(X \in A)$ is defined for all $A \in \mathcal{F}_X$.

For example, in the story that started this chapter, let

N = number of successful emissaries.

Then N is a random variable. Because $N \in \{0, 1, 2, \ldots, 5\}$ with probability 1, the collection of sets that can be assigned probabilities is any subset of $\{0, 1, \ldots, 5\}$.

D24 If $\mathbb{P}(X \in A)$ is defined, say that A is **measurable** with respect to X.

For N,

$\mathcal{F}_N = \{\emptyset, \{0\}, \{1\}, \{2\}, \ldots, \{0, 1, 2, 3, 4, 5\}\}.$

Because there are 6 possible outcomes in $\{0, 1, 2, 3, 4, 5, 6\}$, there are $2^6 = 64$ measurable subsets for N.

The function that takes as input a measurable set $A \in \mathcal{F}_X$ and returns output $\mathbb{P}(X \in A)$ is called the *distribution* of the random variable.

D25 The **distribution** of a random variable X with measurable sets A is the function $\mathbb{P}_X : \mathcal{F}_X \to [0, 1]$ defined as:

$$\mathbb{P}_X(A) = \mathbb{P}(X \in A).$$

Write $X \sim \mathbb{P}_X$.

E24 Suppose $\mathbb{P}(X = 1) = 0.2, \mathbb{P}(X = 2) = \mathbb{P}(X = 3) = 0.4$.
1. What is $\mathbb{P}_X(\{2, 3\})$?
2. What is $\mathbb{P}_X(A)$ for a set A?

Answer. 1. From the information given, $\mathbb{P}_X(\{2, 3\}) = \mathbb{P}(X \in \{2, 3\}) = \mathbb{P}(X = 2) + \mathbb{P}(X = 3) = 0.4 + 0.4 = 0.8$.

2. More generally, for any set A, the probability that X is in A is

$0.2\mathbb{I}(1 \in A) + 0.4\mathbb{I}(2 \in A) + 0.4\mathbb{I}(3 \in A).$

Some distributions occur often enough that it makes sense to give them names. Consider random variables that come from rolling dice.

D26 Write $X \sim \mathrm{d}n$ to mean X is a random variable where $\mathbb{P}(X = i) = 1/n$ for all $i \in \{1, 2, \ldots, n\}$. Say that X is the roll of a fair n-sided die.

The symbol $\sim$ is used to denote the distribution of a random variable. The random variable name goes on the left, and the name of the distribution goes on the right.

F17 If $X \sim \mathrm{d}n$ then

$$\mathbb{P}_X(A) = \sum_{i=1}^{n} \frac{1}{n} \mathbb{I}(i \in A).$$

This works because
$$\mathbb{P}(X \in \{1,\ldots,n\}) = 1.$$

In fact, this sort of construction works for any random variable that only takes on a finite or countable number of values with probability 1. This type of random variable is called *discrete*.

D27 If there exists a sequence $a_1, a_2, \ldots$ such that $\mathbb{P}(X \in \{a_1, a_2, \ldots\}) = 1$, then call X a **discrete** random variable.

F18 For a discrete random variable X with $\mathbb{P}(X \in \{a_1, a_2, \ldots\})$,
$$\mathbb{P}_X(A) = \sum_a \mathbb{P}(X = a)\mathbb{I}(a \in A).$$

So for a discrete random variable X, to determine the distribution of X, it is enough to say what $\mathbb{P}(X = i)$ is for all i such that this value is positive.

SOLVING THE STORY

In the story, each of five emissaries was sent to a neighboring land. Each journey was a success independent of the others with probability 80%.

The letter S can be used to denote a successful journey, while F can be used to denote a failed journey. So

$$SSSFS$$

denotes that journeys 1, 2, 3, and 5 were successful, while journey 4 was a failure. Altogether four journeys were a success so, $N = 4$.

Using independence,
$$\mathbb{P}(SSSFS) = (0.8)(0.8)(0.8)(0.2)(0.8)$$
$$= (0.8)^4(0.2)^1.$$

There are 5 outcomes with $N = 4$:

$$\{FSSSS, SFSSS, SSFSS, SSSFS, SSSSF\}.$$

Each has probability $(0.8)^4(0.2)$ of occurring. Hence
$\mathbb{P}(N = 4) = 5(0.8)^4(0.2) = 0.4096$.

A bit of combinatorics tells us how to find the number of five letter sequences with exactly i letters S and $5 - i$ letters F.

D28 The number of n letter sequences with exactly i letters S and $n-i$ letters F are called the **binomial coefficients**, and are written
$$\binom{n}{i}.$$

The binomial coefficient $\binom{n}{i}$ is read aloud as "n choose i".

F19 The binomial coefficients can be calculated as
$$\binom{n}{i} = \binom{n}{n-i} = \frac{n!}{i!(n-i)!}.$$
(Here $n!$ is the factorial function, where $n! = n(n-1)\cdots 1$ and $0! = 1$.)

For instance, the number of five letter sequences with 3 letters S and 2 letters F are
$$\binom{5}{3} = \frac{5!}{3!2!} = \frac{5 \cdot 4 \cdot 3 \cdot 2 \cdot 1}{3 \cdot 2 \cdot 1 \cdot 2 \cdot 1} = 10.$$

A similar calculation allows determination of all the probabilities of all the possible values of N.

DISTRIBUTION OF N

i	$\mathbb{P}(N = i)$
0	0.00032
1	0.00640
2	0.05120
3	0.20480
4	0.40960
5	0.32768

CHAPTER 7: RANDOM VARIABLES

INDEPENDENCE OF RANDOM VARIABLES

For two events to be independent, the probability that both occurs is the product of the probability that each event occurs.

For two random variables to be independent, for all pairs of measurable sets, the events that the two random variables are in these sets must be independent.

D29 For random variables X and Y to be independent, it must hold for all A measurable with respect to X and B measurable with respect to Y,

$$\mathbb{P}(X \in A, Y \in B) = \mathbb{P}(X \in A)\mathbb{P}(Y \in B).$$

E25 Suppose that D_1, D_2, and D_3 are all independent rolls of a fair six sided die. What is the chance that they are all different numbers?

Answer. First let us write it as several events. Let s_2 be the event that the second die matches the first, and s_3 be the event that the third die matches either the first or the second.

$$s_2 = (D_2 = D_1)$$
$$s_3 = (D_3 = D_1) \vee (D_3 = D_2)$$

Then the goal is to find the probability that neither s_2 nor s_3 occurs. That is, the goal is to find

$$\mathbb{P}(\neg s_2 \wedge \neg s_3) = \mathbb{P}(\neg s_2)\mathbb{P}(\neg s_3 \mid \neg s_2).$$

Since $\neg s_2 = (D_2 \neq D_1)$, then no matter what the value of D_1 is, the chance that the second die roll does not match the first die roll is $5/6$. Hence $\mathbb{P}(\neg s_2) = 5/6$.

Similarly, conditioned on $\neg s_2 = (D_2 \neq D_1)$ being true, for D_3 not to equal either D_1 or D_2 means that it needs to roll one of the four numbers that are left. Hence $\mathbb{P}(\neg s_3 \mid \neg s_2) = 4/6$.

Taken together:

$$\mathbb{P}(\neg s_2 \wedge \neg s_3) = (5/6)(4/6) = 5/9.$$

which is approximately $\boxed{0.5555}$.

Now if $X \in \{1, \ldots, 6\}$ and $Y \in \{1, \ldots, 4\}$, there are 2^6 subsets of $\{1, \ldots, 6\}$ and 2^4 subsets of $\{1, \ldots 4\}$. That means you would have to check $2^{10} = 1024$ pairs of subsets to check independence! The following result cuts down on what needs to be checked considerably.

F20 Suppose $\mathbb{P}(X \in \{x_1, x_2, \ldots\}) = 1$ and $\mathbb{P}(Y \in \{y_1, y_2, \ldots\}) = 1$. Then X and Y are independent if and only if

$$\mathbb{P}(X = x_i, Y = y_j) = \mathbb{P}(X = x_i)\mathbb{P}(Y = y_j)$$

for all i and j.

Proof. One direction is straightforward: if $A = \{x_i\}$ and $B = \{y_j\}$, then independence gives

$$\mathbb{P}(X = x_i, Y = y_j) = \mathbb{P}(X = x_i)\mathbb{P}(Y = y_j)$$

Now suppose that

$$\mathbb{P}(X = x_i, Y = y_j) = \mathbb{P}(X = x_i)\mathbb{P}(Y = y_j)$$

holds for all i and j. Let A and B be arbitrary measurable sets. Then

$$\mathbb{P}(X \in A, Y \in B) = \sum_{x_i \in A, y_j \in B} \mathbb{P}(X = x_i, Y = y_j)$$
$$= \sum_{x_i \in A} \sum_{y_j \in B} \mathbb{P}(X = x_i)\mathbb{P}(Y = y_j)$$

where the sum over x_i and y_j can be broken up into two iterated sums using Tonelli's Theorem. Then

$$\mathbb{P}(X \in A, Y \in B) = \sum_{x_i \in A} \mathbb{P}(X = x_i) \sum_{y_j \in B} \mathbb{P}(Y = y_j)$$
$$= \sum_{x_i \in A} \mathbb{P}(X = x_i) \mathbb{P}(Y \in B)$$
$$= \mathbb{P}(Y \in B) \sum_{x_i \in A} \mathbb{P}(X = x_i)$$
$$= \mathbb{P}(Y \in B) \mathbb{P}(X \in A)$$

as desired. □

For multiple random variables, independence is similar to what was seen before.

D30 A set of random variables $X_1, \ldots, X_n$ are **independent** if for all n-tuples of measurable sets $A_1, \ldots, A_n$

$$\mathbb{P}\left(\bigwedge_{i=1}^{n}(X_i \in A_i)\right) = \prod_{i=1}^{n} \mathbb{P}(X_i \in A_i).$$

D31 A sequence of random variables $X_1, X_2, \ldots$ are **independent** if for all $n \in \{2, 3, \ldots\}$, $X_1, \ldots, X_n$ are independent.

A common situation is when all of a sequence of random variables are both independent and have the same distribution.

D32 A sequence of random variables $X_1, X_2, \ldots$ are **independent and identically distributed** (aka iid) if they are independent, and for all i and j, $X_i \sim X_j$.

σ-ALGEBRAS OF SETS

In this section some of the consequences of the definition of a random variable are explored. Recall that a set of logical statements $\mathcal{F}$ forms a σ-algebra if

1) $(s \in \mathcal{F}) \to (\neg s \in \mathcal{F})$, and

2) $(s_1, s_2, \ldots \in \mathcal{F}) \to (\vee_{i=1}^{\infty} s_i \in \mathcal{F})$.

For a random variable X, $X \in A$ needs to be an event. Recall that $(X \notin A) = (X \in A^C)$, where A^C is the *complement* of A. Also, if there are a sequence of events $X \in A_1, X \in A_2, \ldots$, then saying at least one of these events is true is equivalent to saying that $X \in \cup_{i=1}^{\infty} A_i$. Putting this together:

1) $(X \in A) \in \mathcal{F} \to (X \in A^C) \in \mathcal{F}$.
2) $((X \in A_1), (X \in A_2), \ldots \in \mathcal{F}) \to (X \in \bigcup_{i=1}^{\infty} A_i) \in \mathcal{F}$.

That inspires the definition of a σ-algebra of sets, to avoid having to write $X \in A$ in multiple places.

D33 Say that a nonempty collection $\mathcal{F}$ of sets is a σ-**algebra of sets** if
1. $(\forall A \in \mathcal{F})(A^C \in \mathcal{F})$.
2. $(\forall A_1, A_2, \ldots \in \mathcal{F})(A_1 \cup A_2 \cup \ldots \in \mathcal{F})$.

The point is that no matter what random variable you have, the sets that are measurable with respect to X form a σ-algebra.

F21 For a random variable X, the sets measurable with respect to X form a σ-algebra.

Any σ-algebra of sets contains the empty set.

F22 Any σ-algebra $\mathcal{F}$ contains $\emptyset$.

Proof. A σ-algebra is nonempty, so let $A \in \mathcal{F}$. Then A^C is also in $\mathcal{F}$. Which means

$$[A^C \cup A \cup A \cup A \cup \cdots]^C \in \mathcal{F},$$

but

$$[A^C \cup A \cup A \cup A \cup \cdots]^C = \emptyset,$$

so the result is shown. □

E26 Suppose $\mathbb{P}(X = 1) = 0.2$, $\mathbb{P}(X = 2) = 0.3$, and $\mathbb{P}(X = 3) = 0.5$. Show that $\mathcal{F}_X$ contains all subsets of $\{1, 2, 3\}$.

Answer. It holds that $\emptyset \in \mathcal{F}_X$ since it is always an element of a σ-algebra.

Since $(X = 1) = (X \in \{1\})$ has a defined probability, $\{1\} \in \mathcal{F}_X$. Similarly, $\{2\}$ and $\{3\}$ are measurable. Then

$$\{1\} \cup \{2\} \cup \emptyset \cup \emptyset \cup \cdots \in \mathcal{F}_X$$
$$\{1\} \cup \{3\} \cup \emptyset \cup \emptyset \cup \cdots \in \mathcal{F}_X$$
$$\{1\} \cup \{2\} \cup \{3\} \cup \emptyset \cup \emptyset \cup \cdots \in \mathcal{F}_X,$$

are also measurable, which completes the proof.

Continuous Random Variables

Unlike discrete random variables, which have positive probability of hitting particular values, continuous random variables always have 0 chance of landing on a particular number! In fact, that is how they are defined.

D34 A real-valued random variable X is **continuous** if for all values a, $\mathbb{P}(X = a) = 0$.

How can this happen? Consider trying to draw a point in the interval $(0, 1]$, so all numbers from 0 up to 1 including 1 but not 0.

Begin by flipping a fair coin. Then if it is tails, move to the left half of the interval $(0, 1/2]$ and if heads move to the right half of the interval $(1/2, 1]$. Flip the coin again and repeat, moving either to the left or to the right hand side of the interval as the coin determines.

If an iid sequence of coin flips are used, it can be shown that the chance of landing on a particular number (like 1/3) is 0. This is because with each flip of the coin, the chance that a particular value is landed on is reduced by a factor of $1/2$, and the coin is being flipped an infinite number of times!

The result of this infinite sequence of coin flips is a *standard uniform random variable*, which will be discussed further in the next chapter.

Encounters

79. Ten independent events

Let $X_1, \ldots, X_{10}$ be independent rolls of a fair six-sided die. What is the chance that no 5 is rolled?

80. Different rolls

Let $Y_1, \ldots, Y_{100}$ each be independent rolls of a fair 200-sided die numbered 1 through 200. What is the chance that none of the Y_i equal 7?

81. Flips of a 0-1 coin

Suppose that X_1 and X_2 are random variables that are independent and equally likely to be 0 or 1.

a. If $X_1 = 1$, then what is the probability that $X_2 = 1$?

b. If $X_1 + X_2 \geq 1$ then what is the probability that $X_2 = 1$?

82. Conditioning on sums

Suppose that X_1, X_2, and X_3 are independent, with each equally likely to be 0 or 1.

a. Given that $X_1 + X_2 + X_3 \geq 1$, what is the chance that $X_3 = 1$?

b. Given that $X_1 + X_2 + X_3 \geq 2$, what is the chance that $X_3 = 1$?

83. Six Arrows

An archer shoots six arrows at a target. Each hits (independently of the others) with probability 15%. What is the chance that at least five arrows hit?

84. Ozone

The air quality in a city has a 70% chance of being good, 25% of being medium, and 5% chance of being bad. What is the chance that in the next three days, at least two days are medium air quality?

85. Coin flips

A bored merchant with a coin starts flipping the coin until the first head is seen. If G is the number of flips (including the final one) until a head is seen, then

$$\mathbb{P}(G = i) = (0.2)^{i-1}(0.8),$$

for all $i \in \{1, 2, \ldots\}$.
 Verify that

$$\mathbb{P}(G \in \{1, 2, \ldots\}) = 1.$$

86. Scouting for water

A camp is scouting for water. Each day they send out a scouting party, and there is a 10% chance that they find some. What is the chance that they need at least three days to find water?

87. Measurable sets

Suppose $[3, 4]$, $[4, 5)$, and $[5, 6)$ are measurable with respect to X. Prove that

$$\{[3, 5), [3, 6), [4, 6)\} \subseteq \mathcal{F}_X.$$

88. More measurable sets

Suppose $[-1, 1]$, $[-2, 2]$, $[-3, 3]$, ... are in $\mathcal{F}$. Show that $(-\infty, \infty) \in \mathcal{F}$.

89. The triangle

Suppose that (X, Y) are equally likely to be $(0, 0)$, $(0, 1)$, or $(1, 0)$.
a. What is $\mathbb{P}(X = 0)$?
b. What is $\mathbb{P}(Y = 0)$?
c. What is $\mathbb{P}(X = 0, Y = 0)$?
d. Are X and Y independent?

90. The rectangle

Suppose that (X, Y) has the following distribution.

(x, y)	$\mathbb{P}((X, Y) = (x, y))$
(-1, -1)	0.04
(-1, 1)	0.16
(1, -1)	0.16
(1, 1)	0.64

Show that X and Y are iid.

91. Transport

A bus arrives after a time given by the continuous random variable T.
a. What is $\mathbb{P}(T = 4)$?
b. If $\mathbb{P}(T < 4) = 0.4$, what is $\mathbb{P}(T \leq 4)$?
c. If $\mathbb{P}(T < 4) = 0.4$, what is $\mathbb{P}(T > 4)$?

92. Manufacturing time

Suppose that a product takes random time T to manufacture, and has a number of defects given by N. Say $\mathbb{P}(T < 4, N = 3) = 0.3$, $\mathbb{P}(T < 0.4) = 0.9$ and $\mathbb{P}(N = 3) = 0.4$. Are N and T independent random variables?

Chapter 8: Uniform Random Variables

BUILDING COSTS FOR A NEW castle continued to spiral out of control. The Architect was currently modeling the expenditures as uniform over $[3000, 5000]$ gold pieces. What is the chance the effort would cost at least 4500 gold pieces?

The Discrete Uniform Distribution

Recall that $X \sim \mathrm{d}n$ means that X is the roll of a fair n-sided die. The distributions of the form $\mathrm{d}n$ are a special case of a family of distributions called *uniform*.

> **Uniform means one probability**
> The Latin prefix for one is uni (which is while unicorns have one horn and unicycles have one wheel.) Uniform distributions over $\{a_1, \ldots, a_n\}$ have a single value $1/n$ for the probability that $X = a_i$.

Recall that the counting measure $\#(A)$ of a set A is the number of elements in the set A. With this notation, the general uniform measure over a finite set is as follows.

D35 For a finite set A, say X has **uniform measure** over A if for all a
$$\mathbb{P}(X = a) = \frac{1}{\#(A)}\mathbb{I}(a \in A).$$
Write $X \sim \mathsf{Unif}(A)$.

> **The indicator is essential to the uniform**
> From $\mathbb{I}(a \in A)$ it is possible to find $1/\#(A)$ (that is the normalizing constant) but $\#(A)$ is not enough information to recover $\mathbb{I}(a \in A)$. So it is the $\mathbb{I}(a \in A)$ part that is most important in the uniform distribution!

As a distribution, that is, a function of measurable sets, this looks as follows.

F23 For a finite set A, if $X \sim \mathsf{Unif}(A)$, then
$$\mathbb{P}_X(B) = \mathbb{P}(X \in B) = \frac{\#(AB)}{\#(A)}.$$

Intersection of sets

In the above definition $AB = A \cap B$ refers to the *intersection* of the two sets A and B. The intersection consists of elements that are in both A and B, so it is analogous to logical AND.

D36 The **intersection** of two sets A and B is defined as
$$(x \in A \cap B) = ((x \in A) \wedge (x \in B)).$$
Like logical AND, a comma can be used to indicate intersection of sets, or one can simply write the two sets one after the other. That is,
$$A \cap B = A, B = AB.$$

Since logical AND commutes, so does intersection.

D37 For any two sets A and B,
$$AB = BA.$$

Another useful little fact about intersections that will be used extensively is that the intersection of a set with itself is just the original set.

F24 For any set A, $AA = A$.

Proof. For any logical statement, s, it holds that $ss = s$. Hence for any set A
$$(x \in AA) = (x \in A)(x \in A) = (x \in A),$$
so the sets A and AA are identical. □

Note that this formula looks a lot like our conditional probability formula.

E27 Suppose $X \sim \text{d6}$. Then
$$\mathbb{P}(X \in \{2,5\}) = \frac{\#(\{2,5\})}{\#(\{1,\ldots,6\})} = \frac{2}{6} = \boxed{0.3333\ldots}.$$

F25 For $n \in \{1, 2, \ldots\}$,
$$\text{d}n \sim \text{Unif}(\{1, \ldots, n\}).$$

Consider the following example.

E28 There are 73 animal traps on a factory grounds, of which 4 have animals trapped inside. If a trap is inspected uniformly at random, what is the chance the trap contains a trapped animal?
Answer. Let $A = \{a_1, a_2, a_3, a_4\}$ be the numbers of the traps that contain animals. For $X \sim \text{Unif}(\{1, \ldots, 73\})$, the question is asking what is $\mathbb{P}(X \in A)$? Because X is uniform, and $A \subseteq \{1, \ldots, 73\}$ this is
$$\frac{\#(A)}{\#(\{1,\ldots,73\})} = \frac{4}{73},$$
or about $\boxed{0.05479}$. Only the size of A matters here, not the actual traps that contain the animals.

THE CONTINUOUS UNIFORM DISTRIBUTION

The Discrete Uniform distribution can be thought of as the probability version of counting measure. What if a different measure is used?

Lebesgue measure can be thought of as length in one dimension, area in two dimensions, volume in three dimensions, and so on. Suppose Leb denotes Lebesgue measure the same way $\#$ denotes counting measure. The following are some examples of Lebesgue measure.

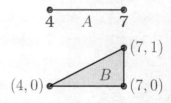

In the above picture, A is a one-dimensional interval, and so the Lebesgue measure is the length of the interval, found by subtracting the smaller endpoint from the greater.
$$\text{Leb}(A) = 7 - 4 = 3.$$

The area B is a triangle, and its Lebesgue measure is the area o the triangle, half of the base times the height.
$$\text{Leb}(B) = (1/2)(3)(1) = 1.5.$$

The Continuous Uniform distribution can be thought of as the probability version of Lebesgue measure, where each set B is assigned probability proportional to the Lebesgue measure of its intersection with A.

D38 For a set A with $\text{Leb}(A) > 0$, say that

$X \sim \text{Unif}(A)$ if

$$\mathbb{P}(X \in B) = \frac{\text{Leb}(AB)}{\text{Leb}(A)}.$$

> **NO UNIFORM DISTRIBUTION OVER POSITIVE INTEGERS**
> There is no such thing as a uniform distribution over the positive integers $\{1, 2, \ldots\}$. The positive integers has infinite counting measure, but Lebesgue measure 0, so it cannot be either a discrete or a continuous uniform!

In one dimension, this reduces to the following.

D39 For nonempty finite interval $[a,b]$, say X has **uniform measure** over $[a,b]$ if for all $a \leq c < d \leq b$,

$$\mathbb{P}(X \in [c,d]) = \frac{d-c}{b-a}.$$

Write $X \sim \text{Unif}([a,b])$.

Note that $b - a$ is the Lebesgue measure of $[a,b]$, and $d - c$ is the Lebesgue measure of $[c,d]$.

THE STANDARD UNIFORM

The continuous uniform over the interval $[0,1]$ is the *standard uniform*.

D40 Let $X \sim \text{Unif}([0,1])$. Then say X is a **standard uniform**.

UNIFORM PROPERTIES

Whether you are working with discrete or continuous uniforms, uniforms all have certain properties in common.

F26 Let $X \sim \text{Unif}(A)$. Then the following hold.
1) $\mathbb{P}(X \in A) = 1$.
2) $\mathbb{P}(X \in B) = \mathbb{P}(X \in A \cap B)$.

In words, these say that the chance that $X \sim \text{Unif}(A)$ falls into A is 1, and that for X to fall into B, then X must fall into A and B.

Proof. Let m be counting measure if A is finite, and Lebesgue measure if $\text{Leb}(A) > 0$.

Then

$$\mathbb{P}(X \in A) = \frac{m(A \cap A)}{m(A)} = \frac{m(A)}{m(A)}$$

For the second part,

$$\mathbb{P}(X \in B) = \frac{m(AB)}{m(A)} = \frac{m(AAB)}{m(A)} = \mathbb{P}(X \in A \cap B).$$

$\square$

Also, conditioning on the value of the uniform leaves a new uniform random variable.

F27 Suppose $X \sim \text{Unif}(A)$. Then

$$[X \mid X \in B] \sim \text{Unif}(A \cap B).$$

Proof. Let A, B, C be measurable sets where $\mathbb{P}(X \in B) > 0$. Then

$$\mathbb{P}(X \in C \mid X \in B) = \frac{\mathbb{P}(X \in C, X \in B)}{\mathbb{P}(X \in B)}$$
$$= \frac{\mathbb{P}(X \in ABC)}{m(AB)/m(A)}$$
$$= \frac{m(ABC)/m(A)}{m(AB)/m(A)}$$
$$= \frac{m(ABC)}{m(AB)},$$

which is the probability a random variable with distribution $\text{Unif}(AB)$ falls into C. $\square$

E29 Suppose that $X \sim \text{d10}$ satisfies $X \leq 8$. What is the distribution of X?
Answer. Note that $\{1, \ldots, 10\} \cap \{1, \ldots, 8\} = \{1, \ldots, 8\}$. So

$$\boxed{[X \mid X \in \{1, \ldots, 8\}] \sim \text{d8}}.$$

Solving the Story

In the story, the cost is

$$C \sim \text{Unif}([3000, 5000]).$$

The question is what is $\mathbb{P}(C \geq 4500)$? From our properties of uniforms, this is

$$\begin{aligned}\mathbb{P}(C \geq 4500) &= \mathbb{P}(C \in [4500, \infty)) \\ &= \mathbb{P}(C \in [4500, \infty) \cap [3000, 5000]) \\ &= \mathbb{P}(C \in [4500, 5000]) \\ &= \frac{5000 - 4500}{5000 - 3000} = \frac{500}{2000},\end{aligned}$$

which is $\boxed{25\%}$.

Encounters

93. Summing three dice

Suppose $D_1 \sim \text{d}10$, $D_2 \sim \text{d}20$ and $D_3 \sim \text{d}6$ are independent. What is

$$\mathbb{P}(D_1 + D_2 + D_3 = 36)?$$

94. More sums of three dice

Suppose $D_1 \sim \text{d}10$, $D_2 \sim \text{d}20$ and $D_3 \sim \text{d}6$ are independent. What is

$$\mathbb{P}(D_1 + D_2 + D_3 = 35)?$$

95. Standard uniform

For $U \sim \text{Unif}([0, 1])$, find:
a. $\mathbb{P}(U \leq 0.7)$.
b. $\mathbb{P}(U \leq -0.7)$.
c. $\mathbb{P}(U \leq 1.3)$.

96. Wider uniform

For $Y \sim \text{Unif}([-1, 1])$, find
a. $\mathbb{P}(Y \leq 0.7)$.
b. $\mathbb{P}(Y \leq -0.7)$.
c. $\mathbb{P}(Y \leq 1.3)$.

97. Functions of a uniform

Prove for the discrete uniform $A \sim \text{Unif}(\{0, 1, 2\})$ that $A^2 \sim \text{Unif}(\{0, 1, 4\})$.

98. Absolutely uniform

Suppose that $T \sim \text{Unif}(\{-2, -1, 1, 2\})$. Prove that $|T| \sim \text{Unif}(\{1, 2\})$.

99. Conditional uniform

Suppose that $Y \sim \text{d}20$. What is the distribution of:
a. Y given that $Y \geq 5$.
b. Y given that $Y < 5$.

100. More conditional uniforms

Suppose $W \sim \text{Unif}([-10, 10])$.
a. What is the distribution of W conditioned on $W \geq 5$?
b. What is the distribution of W conditioned on $W < 5$?

101. Defective testing

In a box with 20 parts, 2 are defective. An inspector picks a part out of the box uniformly at random. What is the probability that the inspector finds a defective part?

102. A deeper test

In a box with 20 parts, 2 are defective. An inspector picks two parts out of the box uniformly at random without replacement. What is the probability that the inspector finds a defective part?

Chapter 9: Functions of Random Variables

THE PARTY OF ADVENTURERS were waiting by the road because they had heard that a mob of bugbears was going to happen by. Soon they grew tired of waiting, and asked the Wizard when they could expect some action. The Wizard sighed, saying that arcane magic had given a model of the bugbears arrival time as an exponential random variable with rate 1.3 per day. They had already waited one day with no results. What is the chance that they would have to wait yet another day?

Functions of Random Variables

When functions are applied to random variables, they change the distribution. For instance suppose that $U \sim \text{Unif}([0,1])$ is a standard uniform random variable. Then

$$\mathbb{P}(U \leq 0.4) = \mathbb{P}(U \in [0, 0.4]) = 0.4 - 0 = 0.4.$$

However,

$$\begin{aligned}\mathbb{P}(U^2 \leq 0.4) &= \mathbb{P}(0 \leq U^2 \leq 0.4) \\ &= \mathbb{P}(0 \leq U \leq \sqrt{0.4}) \\ &= 0.6324\ldots.\end{aligned}$$

So U^2 and U have different distributions.

Now consider $T = -\ln(U)$. That is, T is the negative of the natural logarithm of a standard uniform. Then say that T is a *standard exponential random variable*. To understand how to characterize these distributions, it helps to have a method for determining when two different random variables have the same distribution.

The Cumulative Distribution Function

A distribution of a random variable X is a function that takes any set A and returns the probability that $X \in A$. Checking all such subsets can be very difficult, but it turns out that not every subset probability needs to be checked. The only subsets that need be considered are of the form $(-\infty, a]$. Note that

$$\mathbb{P}(X \in (-\infty, a]) = \mathbb{P}(X \leq a).$$

This motivates the following definition.

D41 The **cumulative distribution function** or **cdf** of a random variable X is defined as

$$\text{cdf}_X(a) = \mathbb{P}(X \leq a).$$

The key reason the cdf is useful is that if two random variables have the same cdf, then they have the same distribution.

F28 If X and Y have the same cdf, then $X \sim Y$.

> This fact is a consequence of an advanced theorem in mathematics called the *Carathéodory extension theorem*, which says that it is possible to extend knowledge of probabilities about sets of the form $(-\infty, a]$ to all possible measurable sets.

Functions keep or destroy information

To understand how functions change random variables, it is important to know that functions can only keep the same amount of information, or possibly less information about their input.

For instance, consider the absolute value function $y = x$. If $x = 3$, then $y = 3$, and if $x = -3$, $y = 3$. So knowing that $y = 3$ has lost some information about x, namely, the sign of x is now unknown.

Not all functions destroy information. One-to-one functions have exactly the same amount of information in the output that was in the input.

Consider a random variable X that is uniform over $[-1, 1]$. Let $Y = X + 3$. Here $f(x) = x + 3$ is a one-to-one function, so $Y = f(X)$ contains exactly the same information as there is in X. Given the value of Y, it is possible to figure out the value of X exactly.

On the other hand, if $W = |X|$, then W contains less information than X did. If $W = 0.3$, then either $X = 0.3$ or $X = -0.3$.

In general, if X is a random variable and f is a function, the information in $f(X)$ is the same or less than the information in X. Consider the following example.

E30 Suppose $X \sim \text{Unif}([-1, 1])$. What is the distribution of $W = |X|$?

Answer. Find the cdf of W. Since absolute value is always nonnegative, if $a < 0$ then $\mathbb{P}(W \leq a) = 0$.

Suppose $a \geq 1$. Then the absolute value of a number in $[-1, 1]$ is at most 1, so $\mathbb{P}(W \leq a) = 1$.

Finally, consider $a \in [0, 1]$.

$$\begin{aligned} \mathbb{P}(W \leq a) &= \mathbb{P}(|X| \leq a) \\ &= \mathbb{P}(-a \leq X \leq a) \\ &= \frac{a - (-a)}{1 - (-1)} \\ &= \frac{2a}{2} = a. \end{aligned}$$

Hence $\text{cdf}_W(a) = a\mathbb{I}(a \in [0, 1]) + \mathbb{I}(a > 1)$. That is the cdf of a standard uniform random variable over $[0, 1]$, so

$$\boxed{W \sim \text{Unif}([0, 1]).}$$

Here is a different function that converts a continuous uniform to a discrete random variable.

E31 Suppose $U \sim \text{Unif}([0, 1])$. Consider

$$W = \mathbb{I}(U \leq 0.3) + 4 \cdot \mathbb{I}(U > 0.3).$$

Find $\mathbb{P}(W = 1)$ and $\mathbb{P}(W = 4)$.

Answer. Since W can only have value 1 or 4, it is a discrete random variable. Here $\mathbb{P}(W = 1) = \mathbb{P}(U \leq 0.3) = 0.3$, and $\mathbb{P}(W = 4) = \mathbb{P}(U > 0.3) = 0.7$, so the distribution is determined by

$$\boxed{\begin{aligned} \mathbb{P}(W = 1) &= 0.3 \\ \mathbb{P}(W = 4) &= 0.7. \end{aligned}}$$

The Bernoulli distribution

In some ways the Bernoulli distribution is the simplest random variable, as it can only take on two values, 0 and 1.

D42 Let $U \sim \mathsf{Unif}([0,1])$. Say that B has the **Bernoulli distribution with parameter** p or is an **indicator random variable**, and write $B \sim \mathsf{Bern}(p)$, if $B = \mathbb{I}(U \leq p)$.

From the definition follows the immediate fact.

F29 For $B \sim \mathsf{Bern}(p)$, $\mathbb{P}(B = 1) = p$ and $\mathbb{P}(B = 0) = 1 - p$.

THE EXPONENTIAL DISTRIBUTION

The natural logarithm of a number in $(0,1]$ is nonpositive. So for $U \in (0,1]$ (which it is with probability 1), $-\ln(U) \in [0, \infty)$. That gives us the following definition.

D43 Let $U \sim \mathsf{Unif}((0,1])$. Then for
$$T = -\ln(U),$$
T has the **standard exponential distribution**.

Often exponential random variables are used to model the time until an event occurs, like in the story at the beginning of the chapter. A parameter called the *rate* can modify the standard exponential distribution. If the rate is higher, that means the event is happening sooner, so the rate *divides* a standard random variable to get the new value.

D44 Suppose W is a standard exponential random variable and $\lambda > 0$. Then $T = W/\lambda$ has an **exponential distribution with rate** λ. Write $T \sim \mathsf{Exp}(\lambda)$.

This is a continuous random variable, since for any a,
$$\mathbb{P}(T = a) = \mathbb{P}(-\ln(U)/\lambda = a)$$
$$= \mathbb{P}(U = \exp(-\lambda a) = 0).$$

The exponential distribution has the following cdf.

F30 If $T \sim \mathsf{Exp}(\lambda)$, then for all a,
$$\mathbb{P}(T \leq a) = [1 - \exp(-\lambda a)]\mathbb{I}(a \geq 0).$$

Proof. Recall that if $U \sim \mathsf{Unif}([0,1])$, then
$$T = -\ln(U)/\lambda$$
has distribution $T \sim \mathsf{Exp}(\lambda)$. Since $\mathbb{P}(U \in (0,1)) = 1$, with probability 1, $T \in [0, \infty)$, and for any $a < 0$,
$$\mathbb{P}(T \leq a) \leq \mathbb{P}(T \leq 0)$$
$$= \mathbb{P}(-\ln(U)/\lambda \leq 0)$$
$$= \mathbb{P}(\ln(U) \geq 0)$$
$$= \mathbb{P}(U \geq 1)$$
$$= 0,$$
which explains the $\mathbb{I}(a \geq 0)$ term on the cdf.

Let $a \geq 0$. Then
$$\mathbb{P}(T \leq a) = \mathbb{P}(-\ln(U)/\lambda \leq a)$$
$$= \mathbb{P}(\ln(U) \geq -\lambda a)$$
$$= \mathbb{P}(U \geq \exp(-\lambda a))$$
$$= 1 - \exp(-\lambda a).$$
□

SOLVING THE STORY PROBLEM

In the story of the wait for the bugbears, the time of the bugbears arrival was being modeled as $T \sim \mathsf{Exp}(1.3/\mathrm{day})$. They had waited one day already, and wondered what was the chance that they would be waiting one day more. In notation, the goal is to find
$$\mathbb{P}(T \geq 2 \mid T \geq 1).$$

By the conditional probability formula, this is

$$\mathbb{P}(T \geq 2 \mid T \geq 1) = \frac{\mathbb{P}(T \geq 2, T \geq 1)}{\mathbb{P}(T \geq 1)}$$
$$= \frac{\mathbb{P}(T \geq 2)}{\mathbb{P}(T \geq 1)}$$
$$= \frac{\exp(-2(1.3))}{\exp(-1(1.3))}$$
$$= \exp(-1.3).$$

Therefore the answer is $\exp(-1.3)$ which is about $\boxed{0.2725}$.

The Fundamental Theorem of Simulation

Any function of $U \sim \mathsf{Unif}([0,1])$ will itself be a random variable. It turns out that this is the *only* way to obtain a real-valued random variable! This result is known as the Fundamental Theorem of Simulation.

T3 The Fundamental Theorem of Simulation. For any real-valued random variable X, there exists a (measurable) function f such that $X = f(U)$, where $U \sim \mathsf{Unif}([0,1])$.

The Self-Replicating Uniform

Recall that if $D_1, D_2, D_3, \ldots$ is an iid stream of $\mathsf{Unif}(\{1, 2, \ldots, 10\})$, then if those numbers are used to form the digits of U, then $U \sim \mathsf{Unif}([0,1])$. For instance, it might be that

$$U = 0.68113356083398457297935\ldots$$

Now suppose instead that two uniforms over $[0,1]$ are desired. Well, just use the odd dice rolls for U_1, and the even dice rolls for U_2'. The result is

$$U_1 = 0.613503947995\ldots$$
$$U_2' = 0.813683852731\ldots$$

Each die roll was independent of the others, so U_1 and U_2 are independent!

Take it further! Use the digits of U_2' to create U_2 and U_3'. Use the digits of U_3' to create U_3 and U_4'. And so on!

The result is an independent, identically distributed stream of uniform random variables

$$U_1, U_2, \ldots.$$

In other words, from a single uniform U over $[0,1]$, it is possible to create an iid stream of random variables $U_1, U_2, \ldots$. Naturally, having an infinite sequence of uniforms gives a lot more flexibility in designing random variables.

D45 Let $U_1, U_2, \ldots$ be iid $\mathsf{Unif}([0,1])$. Then a real-valued random variable is any computable function f applied to $U_1, U_2, \ldots$.

Encounters

103. Finding a CDF

If $X \sim \mathsf{Unif}([-1,2])$, find the cdf of $|X|$.

104. Finding another CDF

If $W \sim \mathsf{Unif}([-2,4])$, find the cdf of W^2.

105. Indicators of a Uniform

If $W \sim \mathsf{Unif}([-2,4])$ and $A = \mathbb{I}(W \leq 3)$, what is the distribution of A?

106. Graphing the CDF

If $R \sim \mathsf{Bern}(1/2)$, plot the cdf of $3R + 1$.

107. Scaling and Shifting Uniforms

Consider $U_1 \sim \text{Unif}([3,8])$ and $U_2 \sim \text{Unif}([0,1])$. Show that U_1 and $5U_2 + 3$ have the same cdf.

108. Exponential Probabilities

For $T \sim \text{Exp}(2)$, find $\mathbb{P}(T \leq 2)$.

109. Scaling Exponentials

Recall that if $U \sim \text{Unif}([0,1])$, $-\ln(U)/\lambda \sim \text{Exp}(\lambda)$. Use this to prove that if $X \sim \text{Exp}(\lambda)$, $X/c \sim \text{Exp}(c\lambda)$ for any nonnegative constant c.

110. A Triple Problem

Suppose $B_1, B_2, \ldots$ are iid $\text{Bern}(0.3)$.
a. What triples $(b_1, b_2, b_3) \in \{0,1\}^3$ satisfy $b_1 \geq \max(b_2, b_3)$?
b. What is $\mathbb{P}(B_1 \geq \max(B_2, B_3))$?

111. Conditional Expectations

Suppose $T \sim \text{Exp}(2.4)$. What is $\mathbb{P}(T \geq 4 \mid T \geq 1)$?

112. What does iid mean?

Suppose $X_1, X_2, X_3, \ldots$ are iid. State whether the following are true for all distributions of (X_1, X_2, X_3) or false if there is at least one distribution of (X_1, X_2, X_3) where the statement is false.
a. X_1, X_2, and X_3 are independent random variables.
b. $\max(X_1, X_2)$, X_1, and X_2 are independent random variables.
c. X_1 and X_{17} have the same distribution.

113. The Ceiling Function

Let $\lceil x \rceil$ be the ceiling function that is the smallest integer greater than or equal to x. So $\lceil 4.3 \rceil = \lceil 5 \rceil = 5$. Note that for an integer i, $\lceil x \rceil = i$ if and only if $i - 1 < x \leq i$. For $U \sim \text{Unif}([0,1])$ find
a. $\mathbb{P}(\lceil 2U \rceil = 2)$
b. $\mathbb{P}(\lceil 2U \rceil = 1)$
c. $\mathbb{P}(\lceil 2U \rceil = 0)$

114. Continuous to Discrete

For $U_1 \sim \text{Unif}((0,1])$, what is the distribution of $N = \lceil nU \rceil$ for a positive integer n?

Chapter 10: The Bernoulli Process

HE HERBALIST WOULD OFTEN venture out into the forest to collect mushrooms. Each trip gave an independent 20% chance of finding a useful bit of fungi. If they went out 10 times, what was the chance that at most three trips were successful?

Stochastic Processes

Whenever you have one or more random variables, you have a *stochastic process*.

D46 A collection of random variables is a **stochastic process**.

One basic type of stochastic process is an *independent, identically distributed* (iid) sequence of random variables.

For instance, consider

$$B_1, B_2, \ldots, \sim \text{Bern}(p),$$

where the $\{B_i\}$ are independent. Recall that $B_i \sim \text{Bern}(p)$ means that $\mathbb{P}(B_i = 1) = p$ and $\mathbb{P}(B_i = 0) = 1 - p$.

This stochastic process is important enough that it gets its own name.

D47 An iid $B_1, B_2, \ldots$ sequence of Bern(p) is called a **Bernoulli process with parameter** p.

Now only consider the i values such that $B_i = 1$. These form a subset of $\{1, 2, \ldots\}$ that can be thought of as a set of points lying within the positive integers. Call this set a *Bernoulli point process*.

D48 A **Bernoulli point process with parameter** p is the subset of positive integers

$$B = \{i : B_i = 1\},$$

where $B_1, B_2, \ldots$ are iid Bern(p).

Because B_i can represent an experiment with two outcomes, often $B_i = 1$ is referred to as a *success*, while $B_i = 0$ is referred to as a *failure*.

E32 Suppose $B_1, B_2, \ldots = 0, 0, 1, 1, 0, 1, \ldots$. Then
$$B = \{3, 4, 6, \ldots\}.$$

It is helpful to have a function that returns the smallest number in B, or ∞ if B is empty. This function is called the *infimum*, and is often abbreviated inf.

D49 For B a subset of $\{1, 2, \ldots\}$, the **infimum** of B is the smallest integer in B if B is nonempty, or ∞ otherwise.

> **INFIMUM**
> More generally, the infimum of a set S is the greatest lower bound on all the elements of S, or $-\infty$ if there is no lower bound on S, or ∞ if S is empty.

There are several questions that can be asked about a Bernoulli point process.

- How many points are in $\{1, \ldots, n\}$? (How many successes were there in the first n trials? What is $B_1 + \cdots + B_n$?)
- How many trials did it take until the first success? (What is $\inf(B)$?)
- More generally, for r a positive integer, how many trials did it take until the rth success? (What is $\inf\{n : B_1 + \cdots + B_n = r\}$?)

Let's answer these questions one at a time.

NUMBER OF SUCCESSES ON n TRIALS

The number of successes in a Bernoulli process in a fixed number of trials is called a *binomial random variable*. Because the B_i are either 0 or 1, to count how many are 1 simply add them together!

> **D50** For a Bernoulli process $B_1, B_2, \ldots$ with parameter p, a binomial random variable with parameters n and p is $B_1 + \cdots + B_n$. Write
> $$B_1 + \cdots + B_n \sim \text{Bin}(n, p).$$

So that is nice as a definition, but how would one (for example) find the chance that out of 5 trials, there were 3 successes?

Using S for success and F for failure, one can write the event
$$B_1 = 1, B_2 = 1, B_3 = 0, B_4 = 1, B_5 = 0$$
much more compactly as
$$SSFSF$$

Because the B_i are independent,
$$\mathbb{P}(SSFSF) = \mathbb{P}(S)\mathbb{P}(S)\mathbb{P}(F)\mathbb{P}(S)\mathbb{P}(F)$$
$$= p^3(1-p)^2.$$

Of course, that is not the only way to get three successes. You could have $SSSFF$ or $FSFSS$ or others! Recall that the number of ways to arrange i letters S and $n-i$ letters F into a length n word is called the *binomial coefficient* n choose i, and can be calculated using
$$\binom{n}{i} = \frac{n!}{i!(n-i)!}.$$

> **THE FACTORIAL FUNCTION**
> Recall that $n!$ (read as n factorial) is the number of 1-1 functions from a domain of size n to a co-domain of size n. To calculate, $n! = \prod_{i=1}^{n} i = n(n-1)(n-2)\cdots 1$. So $4! = (4)(3)(2)(1) = 24$. Also, $0! = 1$.

This gives rise to the following probabilities.

> **F31** For $X \sim \text{Bin}(n, p)$,
> $$\mathbb{P}(X = i) = \binom{n}{i} p^i (1-p)^{n-i} \mathbb{I}(i \in \{0, 1, \ldots, n\}).$$

SOLVING THE STORY

In the story, there were 10 trips into the forest, each with a 20% chance of being successful. Let $T \sim \text{Bin}(10, 0.2)$ be the number of successful trips. Then the goal is to find $\mathbb{P}(T \leq 3)$, which can be broken down as

$$\mathbb{P}(T \leq 3) = \mathbb{P}(T = 0) + \mathbb{P}(T = 1) + \mathbb{P}(T = 2) + \mathbb{P}(T = 3)$$

where

$$\mathbb{P}(T = 0) = \binom{10}{0}(0.2)^0(0.8)^{10}$$
$$\mathbb{P}(T = 1) = \binom{10}{1}(0.2)^1(0.8)^{9}$$
$$\mathbb{P}(T = 2) = \binom{10}{2}(0.2)^2(0.8)^{8}$$
$$\mathbb{P}(T = 3) = \binom{10}{3}(0.2)^3(0.8)^{7}.$$

Summing these values gives a total probability of $\boxed{87.91\%}$.

NUMBER OF TRIALS UNTIL THE 1ST SUCCESS

The number of trials needed until the first success is called a *geometric random variable*.

CHAPTER 10: THE BERNOULLI PROCESS

D51 For $B_1, B_2, \ldots$ iid Bern(p), call
$$G = \inf\{i : B_i = 1\} \sim \text{Geo}(p)$$
a geometric random variable with parameter p.

For example, if $B_1 = 0, B_2 = 1, B_3 = 0, B_4 = 1, \ldots$, then $G = \inf\{2, 4, \ldots\} = 2$. Actually, even knowing B_3 and B_4 were not necessary here: once you see an i with $B_i = 1$ and $B_1 = 0, \ldots, B_{i-1} = 0$, then $G = i$ is the first success.

Now consider $G \sim \text{Geo}(0.2)$. Then what would $\mathbb{P}(G = 3)$ be? Well, in order for G to equal 3, the first two Bernoulli random variables have to be 0, and the third has to be 1. That is,
$$(G = 3) = (B_1 = 0, B_2 = 0, B_3 = 1).$$
Using the independence of the B_i gives
$$\mathbb{P}(G = 3) = \mathbb{P}(B_1 = 0)\mathbb{P}(B_2 = 0)\mathbb{P}(B_3 = 1)$$
$$= (0.8)(0.8)(0.2).$$

In general, the following holds.

D52 For $G \sim \text{Geo}(p)$,
$$\mathbb{P}(G = i) = (1-p)^{i-1} p \cdot \mathbb{I}(i \in \{1, 2, 3, \ldots\}).$$

NUMBER OF TRIALS UNTIL THE rTH SUCCESS

Now jump to a tougher problem: what is the distribution of the number of trials needed to attain the rth success?

For a binomial distribution, the number of trials is fixed and the number of successes is random. Here the number of successes is fixed and the number of trials is random. So this is called the *negative binomial* distribution.

D53 For $B_1, B_2, \ldots$ iid Bern(p), for r a positive integer,
$$N = \inf\{i : B_1 + \cdots + B_i = r\} \sim \text{NegBin}(r, p),$$
and say N is a negative binomial random variable with parameters r and p.

The probabilities for a negative binomial random variable are calculated similarly to a binomial.

F32 For $N \sim \text{NegBin}(r, p)$, $\mathbb{P}(N = i)$ is
$$\binom{i-1}{r-1} p^r (1-p)^{i-r} \cdot \mathbb{I}(i \in \{r, r+1, \ldots\}).$$

Proof. In order for $N = i$, it must be that $B_i = 1$ and $B_1 + \cdots + B_{i-1} = r - 1$. Since these are independent, multiply the probabilities to get that
$$\mathbb{P}(N = i) = \mathbb{P}(B_i = 1)\mathbb{P}(B_1 + \cdots + B_{i-1} = r),$$
so $\mathbb{P}(N = i)$ is
$$p\binom{i-1}{r-1} p^{r-1}(1-p)^{n-1-(r-1)} \mathbb{I}(r-1 \in \{0, \ldots, i-1\}).$$
This can be simplified to give the result. □

THE PICTURE

Consider the following picture, where the points of the Bernoulli process are marked with a red X.

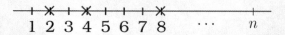

If this draw of the Bernoulli process was being used to make Binomial, Geometric, and Negative Binomial variables, then the results would be as follows.

A Bernoulli process draw

Distribution	In this Bernoulli Process
Bin(3, p)	1
Bin(4, p)	2
Bin(5, p)	2
Geo(p)	2
NegBin(1, p)	2
NegBin(2, p)	4
NegBin(3, p)	8

ENCOUNTERS

115. Basic binomials

Suppose that $X \sim \text{Bin}(100, 0.05)$.
a. What is $\mathbb{P}(X = 0)$?
b. What is $\mathbb{P}(X = 1)$?

116. Tails of binomials

Suppose $N \sim \text{Bin}(6, 0.3)$. What is $\mathbb{P}(N \geq 4)$?

117. Basic geometrics

Let $G \sim \text{Geo}(0.3)$. Find $\mathbb{P}(G \geq 3)$.

118. More geometrics

Suppose $T \sim \text{Geo}(0.8)$.
a. What is $\mathbb{P}(T = 4)$?
b. What is $\mathbb{P}(T \geq 10)$?

119. Negative binomials

Suppose that $R \sim \text{NegBin}(4, 0.25)$.
a. Find $\mathbb{P}(R = 10)$.
b. Find $\mathbb{P}(R \leq 2)$.

120. More negative binomials

Suppose that $T \sim \text{NegBin}(5, 0.3)$.
a. Find $\mathbb{P}(T = 15)$.
b. Find $\mathbb{P}(T \leq 3)$.

121. The drug trial

A drug trial independently brings in 18 patients. Each patient is given a drug expected to lower blood pressure. If the probability the drug works for any given person is 20%, what is the chance that the drug works for at most 5 patients?

122. Waiting for success

With the same drug trial where the chance the drug is effective is 20%, the company decides to alter the experimental protocol as follows. It will keep giving the drug to patients one after another until there are at least 5 successes. What is the chance that they need to see at least 20 patients to make this happen?

123. The bad batch

A company makes about 5,000 of a particular part a year. If there is a 0.001 chance (independently) that each part is a failure, find the probability that there are no failures among the parts.

124. Conditioned on failure

Continuing the last problem, suppose that there is at least one failure among the set of 5,000 parts. What is the chance that there are at least two failures given this information?

125. Fly away

A plane scout looking for forest fires during July in Montana has a 3% chance (independently) of noticing a fire each time a flight is taken. What is the chance that more than 30 flights are needed before the first fire is seen?

126. Looking for medicine

Plants in the Amazon rain forest are being tested for helpful pharmaceutical properties. If any given plant has a 0.0002 chance of showing such properties, what is the chance that looking at 40000 plants finds at least one plant with such properties.

Chapter 11: Poisson point processes

THE TAVERN OWNER MODELED the intervals between customer arrivals as independent random variables with an Exp(6/hr) distribution. What was the chance of getting at least two customers in the first half hour?

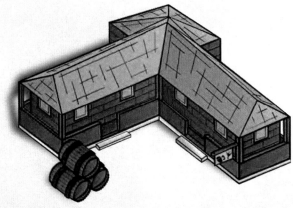

Creating a BPP using Geometrics

Consider a Bernoulli process, $B_1, B_2, \ldots$ iid Bern(p), and a Bernoulli point process
$$B = \{i : B_i = 1\}.$$
Suppose the elements of B are written in order, so
$$B = \{T_1, T_2, \ldots\}$$
where $T_1 < T_2 < T_3 < \cdots$. In that format, the time until the first i with $B_i = 1$ is geometrically distributed. So $T_1 \sim \text{Geo}(p)$. Then the time until the next i with $B_i = 1$ is also geometrically distributed, so $T_2 - T_1 \sim \text{Geo}(p)$.

This idea gives another way to create a Bernoulli point process. Instead of using Bernoulli random variables, use geometric random variables.

F33 Let $G_1, G_2, \ldots$ be iid Geo(p). Then
$$\{G_1, G_1 + G_2, G_1 + G_2 + G_3, \ldots\}$$
forms a Bernoulli point process of rate p.

E33 Suppose $G_1 = 3$, $G_2 = 1$, and $G_3 = 4$. Then the first three points in the Bernoulli point process are 3, $4 = 3 + 1$, and $8 = 4 + 3 + 1$. That is, the point process is
$$P = \{3, 4, 8, \ldots\},$$
and the underlying Bernoulli process $B_1, B_2, \ldots$ equals $0, 0, 1, 1, 0, 0, 0, 1, \ldots$.

Exponential Random Variables

So geometric random variables can be used to create a point process on $\{1, 2, \ldots\}$. What if the goal is to create a point process on $[0, \infty)$? Then a continuous analogue of geometric random variables is needed.

Recall that $G \sim \text{Geo}(p)$ means that for all $i \in \{1, 2, \ldots\}$:
$$\mathbb{P}(G = i) = p(1-p)^{i-1}.$$

Let $\lceil x \rceil$ denote the ceiling of x, which is the function that rounds x up to next integer. For example, $\lceil 4.3 \rceil = 5$, $\lceil 3 \rceil = 3$, $\lceil -2.3 \rceil = -2$. In order for the ceiling of x to equal an integer i, x must be strictly greater than $i - 1$, and at most i. That is,
$$(\lceil x \rceil = i) = (x \in (i-1, i]).$$

That fact gives the following.

F34 Suppose $U \sim \text{Unif}([0,1])$ and $p \in (0,1)$. Then
$$G = \lceil \ln(U)/\ln(1-p) \rceil \sim \text{Geo}(p).$$

Proof. Let $i \in \{1, 2, \ldots\}$ and $\alpha = \ln(1-p) < 0$. Then
$$\mathbb{P}(G = i) = \mathbb{P}(\lceil \ln(U)/\alpha \rceil = i)$$
$$= \mathbb{P}(i - 1 < \ln(U)/\alpha \leq i)$$
$$= \mathbb{P}((i-1)\alpha > \ln(U) \geq i\alpha)$$

The inverse of the natural log function (the exponential function) is increasing, and so preserves the order of inequalities.

$$\mathbb{P}(G = i) = \mathbb{P}(\exp(\alpha(i-1)) > U \geq \exp(\alpha i))$$
$$= \exp(\alpha(i-1)) - \exp(\alpha(i-1))$$
$$= \exp(\alpha(i-1))[1 - \exp(\alpha)].$$

Since $\alpha = \ln(1-p) < 0$, that means
$$\mathbb{P}(G = i) = (1-p)^{i-1}[1-(1-p)] = (1-p)^{i-1}p.$$

□

Note that $-\ln(U) = \ln(1/U)$ has the distribution called the **standard exponential distribution**. Moreover, $-\ln(1/U)/\lambda$ has an exponential distribution with rate λ, and write
$$-\ln(1/U)/\lambda \sim \text{Exp}(\lambda).$$

So the exponential distribution is the continuous analogue of the geometric distribution, with $\lambda = -\ln(1-p)$.

Using Exponentials to Create a Poisson Point Process

Now a continuous analogue to the Bernoulli point process can be built, using exponentials instead of geometrics.

If iid geometrics $G_1, G_2, G_3, \ldots$ with parameter p are summed to give point values, the result is a Bernoulli point process with parameter p.

If iid exponentials $A_1, A_2, \ldots$ with rate λ are summed to give point values, the result is a Poisson point process with rate λ.

D54 Let $A_1, A_2, \ldots$ be iid $\text{Exp}(\lambda)$. Then
$$P = \{A_1, A_1 + A_2, A_1 + A_2 + A_3, \ldots\}$$
is a Poisson point process on $[0, \infty)$ of rate λ. Write $P \sim \text{PPP}(\lambda)$.

Three Questions

With the Bernoulli point process of parameter p, there were three questions to be answered:

1. What is the time until the first point?

2. What is the time until the rth point for some positive integer r?

3. What is the number of points that fall into $\{1, 2, \ldots, n\}$?

The answers (found in the last chapter) are as follows.

F35 Let $B = \{T_1, T_2, \ldots\}$ (with $T_1 < T_2 < \cdots$) be a Bernoulli point process with parameter p over $\{1, 2, \ldots\}$. Then
$$T_1 \sim \text{Geo}(p)$$
$$T_r \sim \text{NegBin}(r, p)$$
$$\#(B \cap \{1, 2, \ldots, n\}) \sim \text{Bin}(n, p).$$

The same three questions can be asked for the Poisson point process, with a slight modification of the last question to deal with continuous state spaces.

CHAPTER 11: POISSON POINT PROCESSES

1. What is the time until the first point?
2. What is the time until the rth point for some positive integer r?
3. What is the number of points that fall into $[0, t]$?

EXPONENTIAL

The answer to the first question is that

$$T_1 \sim \text{Exp}(\lambda),$$

because that is exactly how the Poisson point process over $[0, \infty)$ is defined.

GAMMA

For the second question, recall that an independent exponential is added to T_1 to get T_2, then another independent exponential is added to T_2 to get T_3, and so on. Hence the value of the rth point will be the sum of r iid exponential random variables of rate λ. The distribution of this sum of exponentials is given the name *Gamma*.

> **D55** Suppose $r \in \{1, 2, \ldots\}$ and $A_1, \ldots, A_r$ are iid $\text{Exp}(\lambda)$. Then $T_r = A_1 + A_2 + \cdots + A_r$ has a **gamma distribution with parameters** r **and** λ, and write
>
> $$T_r \sim \text{Gamma}(r, \lambda).$$

> **WRITING OUT THE WORD GAMMA**
> It is customary to write out the word "Gamma" in English when referring to this distribution rather than using the Greek letter. Also, an alternate name for this distribution is *Erlang*, but that is mostly used in Operations Research.

POISSON

Finally consider the number of points that fall into the interval $[0, t]$. This will be a random variable that comes from the *Poisson* family of distributions.

> **D56** Say that a random variable X has the **Poisson distribution with parameter** μ if
>
> $$\mathbb{P}(X = i) = \exp(-\mu)\frac{\mu^i}{i!} \cdot \mathbb{I}(i \in \{0, 1, 2, \ldots\})$$

> **TAYLOR SERIES OF THE EXPONENTIAL FUNCTION**
> Those familiar with the Taylor series expansion of $\exp(x) = 1 + x + x^2/2! + x^3/3! + \cdots$ might note that the probability $X = i$ is the ith (starting with zero) term in the expansion of $\exp(\mu)$. So by multiplying all terms by $\exp(-\mu)$ the probabilities will add up to 1.

The number of points that fall into an interval (a, b) will have a Poisson distribution, with parameter equal to the Lebesgue measure of that interval times the rate of the Poisson point process.

These results can be summarized in the following fact.

> **F36** Let $P = \{T_1, T_2, \ldots\}$ (with $T_1 < T_2 < \cdots$) be a Poisson point process of rate λ over $[0, \infty)$. Then for $a < b$
>
> $$T_1 \sim \text{Exp}(\lambda)$$
> $$T_r \sim \text{Gamma}(r, \lambda)$$
> $$\#(P \cap [a, b]) \sim \text{Pois}((b - a) \cdot \lambda).$$

At this point the tools available are not enough to prove this fact so it will be saved for later.

Unlike the binomial distribution, a Poisson distribution only has one parameter. So the single parameter for the Poisson is the Lebesgue measure of the interval times the rate of the process to give $\text{Pois}(\text{Leb}([a, b])\lambda)$, whereas for the Bernoulli point process the parameters are separate: $\text{Bin}(\#(\{1, \ldots, n\}), p)$.

This last distribution can be used to solve the Tavern story from the beginning of the chapter.

E34 For a Poisson point process of rate 6 per hour, the number of points of the process in $[0, 1/2]$ hours will be Poisson distributed with parameter $(1/2 - 0)6 = 3$. Note the units of hours and per hours cancels out.

The question is asking, then, what is the chance that $X \sim \text{Pois}(3)$ is at least 2? The negation is a bit easier to work with:

$$\mathbb{P}(X \geq 2) = 1 - \mathbb{P}(X \leq 1)$$
$$= 1 - \mathbb{P}(X = 0) - \mathbb{P}(X = 1)$$
$$= 1 - \exp(-3)[1 + 3^1]$$
$$= \boxed{0.8008\ldots}.$$

E35 Suppose $P \sim \text{PPP}(2.1)$. What is $\mathbb{P}(\#(P \cap [2,3]) = 2)$?

Answer. Let $X = \#(P \cap [2,3])$ be the number of points of the Poisson process that fall into the interval $[2, 3]$. Then from the last fact it holds that $X \sim \text{Pois}(2.1 \cdot (3 - 2))$, so

$$\mathbb{P}(X = 2) = \exp(-2.1)\frac{2.1^2}{2!},$$

which is about $\boxed{0.2700}$.

Another useful fact is that the number of points in $[a, b]$ and $[c, d]$ are independent for all $a < b \leq c < d$.

F37 Let $P \sim \text{PPP}(\lambda)$. For $a < b \leq c < d$, let

$$N_{[a,b]} = \#(P \cap [a,b])$$
$$N_{[c,d]} = \#(P \cap [c,d])$$

Then $N_{[a,b]}$ and $N_{[c,d]}$ are independent random variables.

> **OVERLAP AT ONE POINT**
> Note that there can be overlap between the intervals $[a, b]$ and $[c, d]$ of one point. That is because the chance that a particular point will be in the Poisson point process is 0.

ENCOUNTERS

127. TAILS OF EXPONENTIALS

If A is a standard exponential, what is $\mathbb{P}(A \geq 1)$?

128. CONDITIONAL EXPONENTIALS

If A is a standard exponential, what is

$$\mathbb{P}(A \geq 1 \mid A \geq 0.5)?$$

129. BERNOULLIS FROM GEOMETRICS

If $G_1 = 2$, $G_2 = 2$, and $G_3 = 5$ are used to create a Bernoulli point process, what are the values of $B_1, \ldots, B_7$?

130. MORE BERNOULLIS FROM GEOMETRICS

Suppose $G_1 = 3$, $G_2 = 2$, and $G_3 = 1$ are used to create a Bernoulli point process. Then what is $(B_1, B_2, B_3, B_4, B_5)$ for the underlying Bernoulli process?

131. UNDERSTANDING POISSON POINT PROCESSES

Let $P = \{T_1, T_2, \ldots\}$ where $T_1 < T_2 < \cdots$ be a Poisson point process of rate 1.2.
a. What is the distribution of T_1?
b. What is the distribution of T_2?
c. What is the probability that there are exactly two points in $[0, 1]$?

132. FEATURES OF PPP

Let $P = \{T_1, T_2, \ldots\}$ where $T_1 < T_2 < \cdots$ be a Poisson point process of rate 0.7.
a. What is the distribution of T_1?
b. What is the distribution of T_5?
c. What is the probability that there are exactly two points in $[0, 2]$?

CHAPTER 11: POISSON POINT PROCESSES

133. Conditional PPP

Let $P \sim \text{PPP}(1.4)$. What is the probability that there are two points in $[0, 2]$, given that there are four points in $[0, 4]$?

134. More conditional PPP

For a Poisson point process of rate 4.5, what is the probability there is at least one point in $[0, 1]$ given that there are five points in $[0, 10]$?

135. The restaurant

A restaurant models arriving customers as a Poisson point process of rate 70 per hour.

What is the chance that there is at least one customer arrival in the first minute?

136. On the factory floor

A factory models the times of floor stoppages as a Poisson point process with rate 0.01 per day.

What is the chance that no stoppages occur in the first year (365 days)?

Chapter 12: Densities for Continuous Random Variables

The Spring rain was falling heavily in the lake above the dam next to the village. Everyone knew that the dam would eventually fail, but when? The Wizard began to form a model: if T was the time until the dam failed, then the probability that T was in a short interval near t days was about $\exp(-t)\mathbb{I}(t \geq 0)$ times the length of the interval. For example, the chance that $T \in [1.3, 1.31]$ was approximately $\exp(-1.3)(1.31 - 1.3)$. Given this model, what was the chance that the dam would fail during the second day from now? That is, what was $\mathbb{P}(T \in [1,2])$?

Differentials

The chance that a continuous random variable exactly equals a particular value is 0. So how can probabilities be calculated? One way is to use the notion of a *differential*, which intuitively means an infitesimally small change in a variable. For variable x, dx is a differential change in x. For variable t, dt is how the differential is expressed.

A differential can also represent an infitesimally small interval. For instance, use the notation

$$X \in dx,$$

to mean that X is in an interval of length dx that contains the variable value x. That is, (roughly speaking),

$$(X \in dx) = (X \in [x, x + dx]).$$

So depending on context, dx is either the width of a short interval, or the short interval itself. Which one just depends on the context.

In the Story for today, the Wizard's model is

$$\mathbb{P}(T \in dt) = \exp(-t)\, dt.$$

On the left hand side of the equation, the dt is representing an interval, while on the right hand side it is representing the width of that interval.

Now the Wizard really wants to know the probability that $T \in [1, 2]$, not a short interval at all! To get this probability, think about breaking the interval from 1 up to 2 into disjoint little short intervals. Find the probability that T is in each of the short intervals, and add back up to get the probability that T is in the large interval.

The sum of a bunch of infitesimal intervals is called an *integral*, and the sign of an integral is a stretched out letter S for that reason. Overall, this can be written as follows:

$$\mathbb{P}(T \in [1,2]) = \int_{t \in [1,2]} \mathbb{P}(T \in dt).$$

This is a special case of a far-reaching definition.

D57 For a random variable X and measurable set A,
$$\mathbb{P}(X \in A) = \int_{a \in A} \mathbb{P}(T \in da).$$

When the set A is a closed interval $[a, b]$, then an alternate notation puts the a as a subscript of the integral sign, and b as the superscript.

INTEGRAL LIMIT SHORTHAND FOR INTERVALS
The following holds
$$\int_{x \in [a,b]} = \int_a^b$$
for any $a \leq b$.

Returning to the story for a moment, the model gives
$$\mathbb{P}(T \in [1,2]) = \int_1^2 \mathbb{P}(T \in dt)$$
$$= \int_1^2 \exp(-t)\mathbb{I}(t \geq 0) \, dt.$$

Note that the integral is over $t \in [1, 2]$. For each such t value in $[1, 2]$, $\mathbb{I}(t \geq 0) = 1$. So it can be removed in this integral.

$$\mathbb{P}(T \in [1,2]) = \int_1^2 \mathbb{P}(T \in dt) = \int_1^2 \exp(-t) \, dt.$$

More generally, the following rule holds.

F38
$$\int_{x \in A} f(x)\mathbb{I}(x \in B) \, dx = \int_{x \in AB} f(x) \, dx.$$

In this way, an indicator function in the integrand can be moved into the limits of the integral by finding the intersection with the existing limits.

The thing on the right hand side is called a *Lebesgue integral*. This is a generalization of the *Riemann integrals* that are studied in a typical Calculus class. The good news is that if a Riemann integral can be solved to get a number, then the Lebesgue integral has exactly the same value! In this case,

$$\int_1^2 \exp(-t) \, dt = -\exp(-t)|_1^2$$
$$= \exp(-1) - \exp(-2),$$

which is about $\boxed{0.2325}$.

In the problem above, the indicator function did not do much. But for other problems it might have an effect.

E36 Suppose $\mathbb{P}(X \in dx) = (1/2)\mathbb{I}(x \in [0, 2])$. What is $\mathbb{P}(X \geq 1.2)$?
Answer. The indicator function is 0 whenever x is smaller than 0 or greater than 2. This allows us to adjust the limits of our integration.

$$\mathbb{P}(X \geq 1.2) = \int_{x \in [1.2, \infty)} (1/2)\mathbb{I}(x \in [0, 2]) \, dx$$
$$= \int_{x \in [1.2, \infty) \cap [0,2]} 1/2 \, dx$$
$$= \int_{x \in [1.2, 2]} 1/2 \, dx$$
$$= (1/2)(2 - 1.2),$$

which is $\boxed{40\%}$.

DENSITIES

Suppose $\mathbb{P}(X \in da) = f(a) \, da$. Then call $f(a)$ the *density* of the random variable.

D58 If
$$\frac{\mathbb{P}(X \in da)}{da} = f(a),$$
say that the random variable X has **density** $f(a)$. This is also known as the **probability density function**, **pdf**, or **Radon-Nikodym derivative**. If X is a continuous random variable, say that the density is **with respect to Lebesgue measure**.

Often, a subscript indicates which random variable the density is for. For instance, write f_X for the density of X, and f_Y for the density of Y. The name of the function for the density using comes from near the beginning of the alphabet, so f_X, g_X, and h_X are all common choices. An alternative notation for the pdf of a random variable X is pdf_X.

Recall that $\mathbb{P}(X \in (-\infty, \infty)) = 1$ always. It is possible to check that this holds for the two densities that have been seen so far.

$$I = \int_{t \in (-\infty,\infty)} \exp(-t)\mathbb{I}(t \geq 0)\, dt$$
$$= \int_{t \geq 0} \exp(-t)\, dt$$
$$= -\exp(-t)\big|_0^\infty$$
$$= \exp(0) - \lim_{a \to \infty} \exp(-a)$$
$$= 1.$$

For $f_Y(s) = (1/2)\mathbb{I}(s \in [0,2])$,

$$\int_{t \in (-\infty,\infty)} (1/2)\mathbb{I}(t \in [0,2])\, dt = \int_0^2 1/2\, dt$$
$$= (1/2)(2-0)$$
$$= 1.$$

Sometimes a nonnegative function has integral over $(-\infty, \infty)$ that is finite, but the integral is not 1. Such a function is called an *unnormalized density*.

D59 Say that $g \geq 0$ is an **unnormalized density** if

$$0 < \int_{x \in \mathbb{R}} g(x)\, dx < \infty.$$

Call $\left[\int_{s \in \mathbb{R}} g(s)\, ds\right]^{-1}$ the **normalizing constant** and

$$f(x) = \frac{g(x)}{\int_s g(s)\, ds}$$

the normalized density.

E37 What is the normalizing constant for $\exp(-3t)\mathbb{I}(t \geq 0)$?

Solution. To find the normalizing constant, solve the integral!

$$C^{-1} = \int_{t \in (-\infty,\infty)} \exp(-3t)\mathbb{I}(t \geq 0)\, dt$$
$$= \int_{t \in [0,\infty)} \exp(-3t)\, dt$$
$$= \frac{\exp(-3t)}{-3}\bigg|_{t=0}^\infty$$
$$= \frac{1}{3} - \lim_{b \to \infty} \frac{\exp(-3b)}{-3}$$
$$= \frac{1}{3}.$$

Therefore, the normalizing constant is $1/3 = \boxed{0.3333\ldots}$.

The normalized density is

$$\frac{\exp(-3t)\mathbb{I}(t \geq 0)}{1/3} = 3\exp(-3t)\mathbb{I}(t \geq 0).$$

How the CDF and PDF Relate

Often the cdf is known and it would be nice to find the pdf. Or the pdf is known and it would be nice to find the cdf.

Moving from the PDF to the CDF

Recall that for a random variable X, $\text{cdf}_X(a) = \mathbb{P}(X \leq a)$. So to move from a pdf to cdf, just integrate!

F39 For a continuous random variable X with an integrable pdf,

$$\text{cdf}_X(a) = \int_{-\infty}^a \text{pdf}_X(r)\, dr.$$

E38 Suppose Y has $\text{pdf}_Y(s) = 3\exp(-3s)\mathbb{I}(s \geq 0)$. Find the cdf of

Y.

Answer. For $a < 0$, the answer I_- is

$$I_- = \int_{s \in (-\infty, a]} 3\exp(-3s)\mathbb{I}(s \geq 0)\, ds$$
$$= \int_{s \in (-\infty, a] \cap [0, \infty))} 3\exp(-3s)\, ds$$
$$= \int_{s \in \emptyset} 3\exp(-3s)\, ds = 0.$$

For $a \geq 0$, the answer I_+ is

$$I_+ = \int_{s \in (-\infty, a]} 3\exp(-3s)\mathbb{I}(s \geq 0)\, ds$$
$$= \int_{s \in (-\infty, a] \cap [0, \infty))} 3\exp(-3s)\, ds$$
$$= \int_{s \in [0, a]} 3\exp(-3s)\, ds$$
$$= -\exp(-3s)\big|_0^a = 1 - \exp(-3a).$$

Hence the solution is

$$\boxed{(1 - \exp(-3a))\mathbb{I}(a \geq 0).}$$

MOVING FROM THE CDF TO THE PDF

Given a cdf, what is $\mathbb{P}(X \in (a, a + da])$? Well, $(-\infty, a])$ and $(a, a + da]$ are disjoint intervals whose union is $(-\infty, a + da])$. Therefore,

$$\text{cdf}_X(a) + \mathbb{P}(X \in da) = \text{cdf}_X(a + da),$$

or rearranging

$$\mathbb{P}(X \in da) = \text{cdf}_X(a + da) - \text{cdf}_X(a).$$

Dividing both sides by da gives

$$\text{pdf}_X(a) = \frac{\mathbb{P}(X \in da)}{da}$$
$$= \frac{\text{cdf}_X(a + da) - \text{cdf}_X(a)}{da}.$$

The right hand side should look familiar, this is the *derivative* of the cdf of X. This motivates (because it is not a formal proof of) the following fact.

F40 If X has a differentiable cdf, then

$$\text{pdf}_X(a) = [\text{cdf}_X(a)]',$$

at all but a countable number of values of a.

The key to differentiating an expression with indicator functions is that when the indicator function is 0, the derivative is 0. When the indicator function is 1, the derivative is the derivative of whatever function multiplies the indicator function. In other words, indicator functions can be brought out of the derivative operator.

F41 If $f(x)$ is a differentiable function, then

$$[f(x)\mathbb{I}(x \in A)]' = \mathbb{I}(x \in A)f'(x).$$

E39 Suppose that Y has cdf

$$\text{cdf}_Y(a) = a\mathbb{I}(a \in [0, 1]) + \mathbb{I}(a > 1).$$

Find pdf_Y.

Answer. This cdf is differentiable everywhere except at $a = 0$ and $a = 1$. The graph looks like this:

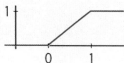

Differentiating using the rule for indicator functions gives:

$$\text{pdf}_Y(a) = [\text{cdf}_Y(a)]'$$
$$= [a\mathbb{I}(a \in [0, 1]) + 1 \cdot \mathbb{I}(a > 1)]'$$
$$= [a]'\mathbb{I}(a \in [0, 1]) + [1]'\mathbb{I}(a > 1)$$
$$= (1)\mathbb{I}(a \in [0, 1]) + (0)\mathbb{I}(a > 1).$$

Hence

$$\boxed{\mathbb{I}(a \in [0, 1])}$$

is the pdf of Y.

Encounters

137. Density of a Uniform

Suppose $\mathbb{P}(Y \in dy) = (1/30)\mathbb{I}(y \in [0, 30])\, dy$. What is $\mathbb{P}(Y \leq 5)$?

138. Gamma density

Suppose $\mathbb{P}(Y \in dy) = 4y \exp(-2y)\mathbb{I}(y \geq 0)\, dy$. What is $\mathbb{P}(Y \geq 1)$?

139. More uniform density

Say that $\mathbb{P}(W \in dw) = (1/20)\mathbb{I}(w \in [30, 50])\, dw$. What is the density of W?

140. Exponential density

Say that $\mathbb{P}(T \in dt) = \exp(-t)\mathbb{I}(t \geq 0)\, dt$. What is the density of T?

141. Another exponential density

Suppose $\mathbb{P}(R \in dr) = 3\exp(-3r)\mathbb{I}(r \geq 0)\, dr$. Find $\text{cdf}_R(a)$.

142. Cdf of a gamma

Let $\text{pdf}_Y(y) = 4y\exp(-2y)\mathbb{I}(y \geq 0)$. Find $\text{cdf}_Y(a)$.

143. Normalizing an exponential

Suppose $\mathbb{P}(R \in dr) = C\exp(-3r)\mathbb{I}(r \in [0, 2])\, dr$. Find C.

144. Normalizing a beta density

Suppose $\mathbb{P}(A \in da) = Ca^2\mathbb{I}(a \in [0, 1])\, da$. Find C.

145. Density of a function of a uniform

Suppose $U \sim \text{Unif}([0, 1])$, and $W = \sqrt{U}$.
a. Find cdf_W.
b. Find pdf_W.

146. Shifting and scaling a uniform

Suppose $U \sim \text{Unif}([0, 1])$, and $S = 3U + 4$.
a. Find cdf_S.
b. Find pdf_S.

147. From cdf to pdf

Suppose $\text{cdf}_X(x) = (1 - 1/x)\mathbb{I}(x \geq 1)$. Find $\text{pdf}_X(x)$.

148. Double exponential

Suppose X has cdf

$$\text{cdf}_X(x) = (1/2)e^a\mathbb{I}(a < 0) + (1 - (1/2)e^{-a})\mathbb{I}(a \geq 0).$$

Find $\text{pdf}_X(x)$.

CHAPTER 12: DENSITIES FOR CONTINUOUS RANDOM VARIABLES

Chapter 13: Densities for Discrete Random Variables

As the army approached the gates, the King turned to his Scouts and asked: "How large is the approaching menace?" The Scouts huddled together, and reported: "We are not sure your majesty. However, we can say that the probability they number n thousand troops is $3^n \exp(-3)/n!$."

The King sighed, wondering why he had hired scouts so invested in probabilistic modeling. "Just tell me", he pleaded, "can you give me the number that is most likely?"

Densities against counting measure

Densities for continuous random variables are with respect to Lebesgue measure, which is why to find $\mathbb{P}(X \in A)$ when X is continuous, a Lebesgue integral is used.

Discrete random variables also have densities, but they are with respect to *counting measure* rather than Lebesgue measure.

For counting measure, think about what a little differential element $[x - dx/2, x + dx/2]$ means. Consider y. The only way for y to fall in the interval $[x - dx/2, x + dx/2]$ is if $y = x$, in which case a count of 1 is registered. Otherwise, if $y \neq x$, a count of 0 is registered.

The result is that

$$\mathbb{P}(X \in da) = \mathbb{P}(X = a)$$

when working with discrete random variables.

Now consider how integration is performed against counting measure. For instance, consider

$$\int_1^5 x^2 \mathbb{I}(x \in \{1, 2, 3, 4, 5\}) \, d\#.$$

There are only five values of x where the differential is positive, $x = 1, \ldots, x = 5$. The first value $x = 1$, has $d\# = 1$ so the integral gets a contribution of $1^2 = 1$. The second value $x = 2$, also has $d\# = 1$ again so the integral gets a contribution of $2^2 = 4$. Working towards 5 gives

$$S = \int_1^5 x^2 \mathbb{I}(x \in \{1, 2, 3, 4, 5\}) \, d\#$$
$$= \sum_{i=1}^{5} i^2$$
$$= 1^2 + 2^2 + 3^2 + 4^2 + 5^2 = 55.$$

In general, integrals with respect to counting measure are a sum.

F42 Let $\#$ denote counting measure. Then for a countable set A,

$$\int_{a \in A} f(a) \, d\# = \sum_{a \in A} f(a).$$

E40 Let $X \sim d6$ be the roll of a fair six-sided die. Then the density of X with respect to counting measure is

$$\text{pdf}_X(i) = (1/6) \mathbb{I}(i \in \{1, 2, 3, 4, 5, 6\}).$$

What is $\mathbb{P}(X \leq 2)$?

Answer. Note $\mathbb{P}(X \leq 2) = \mathbb{P}(X \in \{1,2\})$, which is

$$\mathbb{P}(X \in \{1,2\}) = \int_{i \in \{1,2\}} \mathrm{pdf}_X(i)\, d\#$$
$$= \sum_{i=1}^{2} \frac{1}{6} \mathbb{I}(i \in \{1,\ldots,6\})$$
$$= \frac{1}{6} + \frac{1}{6} = \frac{1}{3},$$

which is about $\boxed{0.3333}$.

> **THE PROBABILITY MASS FUNCTION**
> An old term for the density of a discrete random variable is *probability mass function* which is unfortunately a terrible name that makes absolutely no sense because mass can either be concentrated in chunks or spread out continuously like peanut butter. Still, the term is used now and then, so be aware that the density of discrete random variables is sometimes called the pmf instead of the pdf.

D60 For a discrete random variable X, the **density** (aka **pdf** aka **probability mass function** aka **pmf**) is

$$\mathrm{pdf}_X(i) = \mathbb{P}(X = i).$$

DISCRETE CDF FUNCTIONS

Turning a pdf into a cdf is the same for the discrete world as the continuous. The only thing to note is that the integral becomes a sum.

$$\mathbb{P}(X \leq a) = \int_{-\infty}^{a} \mathrm{pdf}_X(s)\, d\# = \sum_{i \leq a} \mathrm{pdf}_X(i).$$

The fact that the cdf is a sum leads to the cdf having *jumps*. That is, the cdf is continuous for continuous random variables, and has jumps when the random variable is discrete.

E41 Draw the cdf of X when $\mathbb{P}(X = 1) = 0.2$, $\mathbb{P}(X = 3) = 0.5$, $\mathbb{P}(X = 5) = 0.3$.

Answer. The cdf of X will have a jump of size 0.2 at $x = 1$, a jump of size 0.5 at $x = 3$, and a final jump of size 0.3 at $x = 5$.

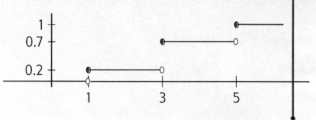

MEASURES OF CENTRAL TENDENCY

For a random variable with density f, where is the center of the distribution? There is more than one way to measure that.

THE MODE

D61 Say that a random variable X with density f_X has mode m if

$$m \in \arg\max f_X(x).$$

The set of all modes of X is the **mode set** of the random variable.

If X is a discrete random variable, a mode x has

$$\mathbb{P}(X = x) \geq \mathbb{P}(X = s)$$

for all other values s.

If X is a continuous random variable, then a mode is any value x such that $\mathbb{P}(X \in dx) \geq \mathbb{P}(X \in ds)$ for all other s. Note that the mode might not be unique.

For discrete random variables that are integer valued with probability 1, a way to find the mode is to look at *ratios* of successive values of the pdf.

For N such that $\mathbb{P}(N \in \{1, 2, 3, \ldots\}) = 1$, let
$$r(i) = \frac{\operatorname{pdf}_N(i+1)}{\operatorname{pdf}_N(i)}.$$

Note that if $r(i) > 1$, then $\operatorname{pdf}_N(i+1) > \operatorname{pdf}_N(i)$, so that eliminates i as a possible mode of N.

If $r(i) < 1$, then $\operatorname{pdf}_N(i+1) < \operatorname{pdf}_N(i)$, so that eliminates $i+1$ as a possible mode of N. If $r(i) = 1$, then $\operatorname{pdf}_N(i+1) = \operatorname{pdf}_N(i)$, so both might be modes of N.

Typically this can be used to eliminate most integers from contention for mode. Since an integer valued random variable has to have at least one mode, usually whatever is not eliminated is the mode. Often there is only one mode or two modes right next to each other. This is the basis of the following fact.

F43 Consider an integer valued random variable N with
$$r(i) = \frac{\operatorname{pdf}_N(i+1)}{\operatorname{pdf}_N(i)}.$$

Suppose there is a noninteger a such that $r(i) > 1$ for $i < a$ and $r(i) < 1$ for $i > a$. Then $\{\operatorname{floor} a + 1\}$ is the mode set of N.

If there is an integer a such that $r(i) > 1$ for $i < a$, $r(i) < 1$ for $i > a$, and $r(a) = 1$, then the mode set is $\{a, a+1\}$.

E42 In the Story for today, $\operatorname{pdf}_N(n) = \exp(-3) 3^n / n!$, and the goal is to find the mode. That is, what value of n makes $\operatorname{pdf}_N(n)$ as large as possible?
Answer. To find out, consider the ratio
$$r(n) = \frac{\operatorname{pdf}_N(n+1)}{\operatorname{pdf}_N(n)} = \frac{\exp(-3) 3^{n+1}/(n+1)!}{\exp(-3) 3^n/n!} = \frac{3}{n+1}.$$

In the problem $r(n)$ is greater than 1 if $n + 1 < 3 \to n < 2$, and less than 1 if $n + 1 > 3 \to n > 2$. When $n = 2$ it holds that $r(2) = 1$.

Hence $\boxed{n = 2 \text{ and } n = 3}$ are both modes of N.

The ratio method works exceedingly well when there are $n!$ factors in the pdf, as they mostly cancel out as seen above. For continuous pdfs, derivatives are usually the way to go.

E43 Suppose $\operatorname{pdf}_X(s) = s^2 \exp(-3s) \mathbb{I}(s \geq 0)$. What is the mode(s) of X?
Answer. When $s \leq 0$, the derivative of the pdf of X is 0. Now consider $[\operatorname{pdf}_X(s)]'$ when $s \geq 0$. Then using the chain rule:
$$[\operatorname{pdf}_X(s)]' = [s^2 \exp(-3s)]'$$
$$= ([s^2]' \exp(-3s) + s^2 [\exp(-3s)]')$$
$$= (2s \exp(-3s) + s^2(-3) \exp(-3s))$$
$$= \exp(-3s) s (2 - 3s).$$

So the derivative of the density for all s is
$$[\operatorname{pdf}_X(s)]' = \exp(-3s) s (2 - 3s) \mathbb{I}(s \geq 0).$$

No matter what s is, $\exp(-3s) > 0$. Also, $s\mathbb{I}(s \geq 0) \geq 0$ for all s. So when $s > 0$ and $2 - 3s < 0$ the derivative is negative, and when $s > 0$ and $2 - 3s > 0$ the derivative is positive. This switch occurs at $s = 2/3$, so this is where the mode occurs, at approximately $\boxed{0.6666}$.

THE MEDIAN

Another way to look at the center is where the cdf and one minus the cdf are both at least $1/2$.

D62 Say that m is a **median** of a random variable X if both $\mathbb{P}(X \leq m) \geq 1/2$ and $\mathbb{P}(X \geq m) \geq 1/2$. The set of all medians of a random variable X is the **median set**.

E44 Suppose $A \sim \mathrm{d}5$. What is the median set of A?
Answer. For $i \in \{1, 2, 3, 4, 5\}$, $\mathbb{P}(A \leq i) = i/5$. So any median must have $i \geq 3$. Similarly, $\mathbb{P}(A \geq i) = (6 - i)/5$. So any median must have $i \leq 3$. The only number that fits the bill is $\boxed{3}$.

For discrete random variables, the median set can contain more than one value.

> **E45** For $B \sim $ d6, what is the median set of B?
> **Answer.** For $a \in [1,6]$, $\mathbb{P}(B \leq a) = \lfloor a \rfloor/6$. So any median must have $i \geq 3$. Similarly, $\mathbb{P}(B \geq a) = \lfloor 7-i \rfloor/6$. So any median must have $i \leq 4$.
>
> On the other hand, if $i \in [3,4]$, then $\mathbb{P}(X \leq i) \geq \mathbb{P}(X \leq 3) = 1/2$, and $\mathbb{P}(X \geq i) \geq \mathbb{P}(X \geq 4) = 1/2$. Therefore, the median set is $\boxed{[3,4]}$.

SHIFTING AND SCALING

When a constant is subtracted from a random variable, that is called *shifting* the random variable. Multiplying a random variable by a constant is called *scaling* the random variable. How that affects the density depends on whether it is applied to a continuous random variable or a discrete random variable.

> **F44** Let $a \neq 0$, $b \in \mathbb{R}$, X be a random variable with pdf f_X, and $Y = aX + b$. If X is a continuous random variable, then
> $$f_Y(s) = \frac{1}{|a|} f_X((s-b)/a).$$
> If X is a discrete random variable, then
> $$f_Y(s) = f_X((s-b)/a).$$

Why the difference between continuous and discrete? Because in the continuous world, for $y = ax + b$, $dy = a\,dx$, while for the discrete world dy and dx are equal. Or you can think about the proof.

Proof. Let X be continuous. Let $a > 0$, then by the chain rule

$$\begin{aligned} f_Y(s) &= [\mathbb{P}(aX + b \leq s)]' \\ &= [\mathbb{P}(X \leq (s-b)/a)]' \\ &= f_X((s-b)/a)[(s-b)/a]' \\ &= f_X((s-b)/a)(1/a). \end{aligned}$$

When $a < 0$,

$$\begin{aligned} f_Y(s) &= [\mathbb{P}(aX + b \leq s)]' \\ &= [\mathbb{P}(X \geq (s-b)/a)]' \\ &= [1 - \mathbb{P}(X < (s-b)/a)]' \\ &= -f_X((s-b)/a)[(s-b)/a]' \\ &= f_X((s-b)/a)(1/(-a)). \end{aligned}$$

So in both cases

$$f_Y(s) = f_X((s-b)/a)/|a|.$$

For discrete random variables

$$\begin{aligned} f_Y(i) &= \mathbb{P}(Y = i) \\ &= \mathbb{P}(aX + b = i) \\ &= \mathbb{P}(X = (i-b)/a) \\ &= f_X((i-b)/a). \end{aligned}$$

$\square$

> **E46** Suppose $T \sim \text{Exp}(1)$, what is the density of $3T + 4$?
> **Answer.** The density of T is $f_T(s) = \exp(-s)\mathbb{I}(s \geq 0)$. Plugging in $s = (t-4)/3$ gives
> $$\begin{aligned} f_{3T+4}(t) &= \frac{1}{|3|} f_T((t-4)/3) \\ &= \frac{1}{3} \exp\left(-\frac{t-4}{3}\right) \mathbb{I}\left(\frac{t-4}{3} \geq 0\right), \end{aligned}$$
> which simplifies to
> $$\boxed{\frac{\exp(4/3)}{3} \exp(-t/3)\mathbb{I}(t \geq 4).}$$

> **E47** Suppose $X \sim $ d3. What is the density of

$-2X$?

Answer. Since X is a discrete random variable,

$$f_{-2X}(i) = f_X(i/(-2))$$
$$= \frac{\mathbb{I}(i/(-2) \in \{1,2,3\})}{3},$$

and solving for i in the indicator function gives

$$\boxed{f_{-2X}(i) = \frac{\mathbb{I}(i \in \{-2,-4,-6\})}{3}.}$$

THE SURVIVAL FUNCTION

If an item lasts T time before breaking, then if $T > t$ the item has survived past time t. This motivates the definition of the *survival function*.

D63 The survival function of a random variable X is

$$\mathrm{sur}_X(t) = 1 - \mathrm{cdf}_X(t) = \mathbb{P}(X > t).$$

ENCOUNTERS

149. PDF FOR A DIE

If $Y \sim \mathrm{d}10$, what is $\mathrm{pdf}_Y(i)$?

150. CDF FOR A DIE

For $X \sim \mathrm{d}6$, draw $\mathrm{cdf}_X(a)$.

151. BINOMIAL MODE

Suppose $\mathrm{pdf}_B(i) = \binom{10}{i} p^i (1-p)^{10-i}$.
a. What is $\mathbb{P}(B = 6)$?
b. What is the mode if $p = 0.42$?

152. GAMMA MODE

Suppose $\mathrm{pdf}_G(r) = (1/6) r^3 \exp(-r) \mathbb{I}(r \geq 0)$. Find the mode of G.

153. GAMMA MEDIAN

Suppose $\mathrm{pdf}_G(r) = (1/6) r^3 \exp(-r) \mathbb{I}(r \geq 0)$. Find the median of G.

154. MEDIAN SETS OF A DIE

For $X \sim \mathrm{d}10$, find the median set of X.

155. CDF OF A SCALED DIE

For $X \sim \mathrm{d}6$, what is the cdf of $2X$?

156. PDF OF A SCALED DIE

For $X \sim \mathrm{d}6$, what is the pdf of $2X$?

157. SCALING A BETA

For T with $\mathrm{pdf}_T(t) = 12 t^2 (1-t) \mathbb{I}(t \in [0,1])$, what is the pdf of $2T$?

158. SCALED AND SHIFTED GAMMA

For G with pdf $(1/6) r^3 \exp(-r) \mathbb{I}(r \geq 0)$, what is the pdf of $3G + 1$?

159. SURVIVAL FUNCTION OF AN EXPONENTIAL

For $T \sim 4 \exp(-4t) \mathbb{I}(t \geq 0)$, what is the survival function of T?

160. SURVIVAL FUNCTION OF A DIE

For $X \sim \mathrm{d}6$, what is the survival function of X?

Chapter 14: Mean of a Random Variable

The Elder knew that the taxes collected from the members of the village varied widely. About 10% of the villagers could contribute 5 florins a year, while 40% could contribute 2 florins, and the remaining 50% only 1. What, the Elder wondered, was the average amount of taxes that a villager paid?

The Sample Average

Consider a random variable X that takes on different values. The variable X could be like the taxes paid by villagers in the story, where

$$\mathbb{P}(X = 5) = 0.1$$
$$\mathbb{P}(X = 2) = 0.4$$
$$\mathbb{P}(X = 1) = 0.5.$$

Suppose that there were a 1000 villagers, paying taxes $X_1, \ldots, X_{1000}$. The sample average of the taxes paid by the villagers would be

$$\frac{X_1 + X_2 + \cdots + X_{1000}}{1000}.$$

Because of the probabilities assigned to each value of X, it is not unreasonable to believe that there would be about 100 that paid 5, 400 that paid 2, and 500 that paid 1. Of course, because the random variables are random, those would not be exactly the values, but close.

If these were the exact numbers, say that $x_1 = x_2 = \ldots = x_{100} = 5$, $x_{101} = \ldots = x_{500} = 2$, and $x_{501} = \ldots = x_{1000} = 1$. Then the sample average of x would be:

$$\frac{5 + \cdots + 5 + 2 + \cdots + 2 + 1 + \cdots + 1}{1000},$$

which makes

$$\frac{(5)(100) + (2)(400) + (1)(500)}{1000},$$

and dividing the 1000 through the numerator gives

$$5(0.1) + 2(0.4) + 1(0.5).$$

This number is a *weighted average* of the possible outcomes. Each outcome is multiplied by the probability of that outcome, and the whole thing is then summed up.

Define this number to be the *expected value* or *mean* of the random variable. Like the median, the mean is another example of a *measure of central tendency*, and only depends on the distribution of the random variable.

D64 Suppose $\mathbb{P}(X \in \{x_1, \ldots, x_n\}) = 1$. Then the **expected value**, also known as the **expectation**, **average**, or **mean** of the random variable, is

$$\sum_{i=1}^{n} x_i \mathbb{P}(X = x_i).$$

When $U \sim \mathsf{Unif}(\{u_1, \ldots, u_n\})$, then

$$\mathbb{E}[U] = \sum_{i=1}^{n} \frac{1}{n} u_i$$

is just the sample average of the u_i.

Indicator functions have especially nice means that connect averages with probabilities.

F45 For s an event,
$$\mathbb{E}[\mathbb{I}(s)] = \mathbb{P}(s).$$

Proof. Since $\mathbb{I}(s)$ is either 0 or 1,
$$\mathbb{E}[\mathbb{I}(s)] = (0)\mathbb{P}(\mathbb{I}(s)=0)+(1)\mathbb{P}(\mathbb{I}(s)=1) = \mathbb{P}(s).$$
□

So really, all of probability can be considered a special case of expected value!

MOVING TO THE INFINITE

The definition given works for when X only takes on a finite number of values, but what about distributions like the geometric, where the values could be $\{1, 2, \ldots\}$? Then an infinite sum is needed. Unfortunately, not all infinite sums converge. For those that do, the random variable is called *integrable*.

D65 Let X be a random variable with $\mathbb{P}(X \in \{x_1, x_2, \ldots\}) = 1$. Then consider the infinite sum
$$\sum_{i=1}^{\infty} x_i \mathbb{P}(X = x_i).$$
If this sum converges, call the result the expected value of X, and say that X is integrable.

Note that if $W \geq 0$ with probability 1, then $\sum_{i=1}^{\infty} w_i \mathbb{P}(W = w_i)$ will always be either a real number or ∞. This leads to the following fact.

F46 A random variable W is integrable if and only if $\mathbb{E}[|W|] < \infty$.

Of course, if a random variable X only takes on a finite number of values, it is always integrable.

F47 If $\mathbb{P}(X \in \{x_1, \ldots, x_n\}) = 1$, then X is integrable.

THE STRONG LAW OF LARGE NUMBERS

So now the expected value is defined for a discrete random variable, but how does that relate to the sample average from the story?

For instance, if $X_1, \ldots, X_{1000}$ are iid draws with the same distribution as X, and X has mean $\mathbb{E}[X]$, how does the sample average
$$\frac{X_1 + \cdots + X_{1000}}{1000}$$
and $\mathbb{E}[X]$ compare? Will they be close together?

It turns out the answer to that question is yes, and is the subject of one of the most important theorems in probability, the Strong Law of Large Numbers.

T4 The Strong Law of Large Numbers. Given a sequence of random variables $X_1, X_2, \ldots$ that are iid with distribution identical to X where $\mathbb{E}[X] < \infty$. Then
$$\mathbb{P}\left(\lim_{n \to \infty} \frac{X_1 + X_2 + \cdots + X_n}{n} = \mathbb{E}[X]\right) = 1.$$

There is a lot to unpack there! Here are some of the highlights

- Because $X_1, X_2, \ldots$ are random variables, so are $S_n = (X_1 + \cdots + X_n)/n$ for any n.
- The random variables $X_1, X_2, \ldots$ are iid, but the sample averages S_n are not independent, since S_{n+1} depends on the value of S_n.
- Even though the S_n are not independent, it is possible to say something about how they behave.

With probability 1, they will get closer and closer to $\mathbb{E}[X]$ if X is an integrable random variable.
- If $\mathbb{E}[X]$ does not exist, it turns out the limit will never exist with probability 1. (The limit never existing includes when the limit is ∞.)

EVENTS OF PROBABILITY 1

The "with probability 1" part in the SLLN seems a bit weird. Recall that events that are always true have probability 1. However, the reverse is not true! When dealing with an infinite number of possible outcomes, it is possible to have an event which is not logically true, but still has probability 1 of being true.

It helps to look at an example. Suppose $D \sim \text{Unif}(\{1,2,3,4,5,6\})$. Then

$$\mathbb{E}[D] = (1/6)(1 + \cdots + 6) = 3.5.$$

Hence, for $D_1, D_2, D_3, \ldots$ iid d6, the sample average will converge to 3.5.

But what if $D_1 = D_2 = D_3 = \cdots = 2$? Then every sample average, $S_n = (D_1 + \cdots + D_n)/n$ equals 2 as well. So the limit of the sample averages is 2, not 3.5! It is *possible* for the limit to not equal 2. But what is the probability?

The chance that every single die roll is 2 is

$$(1/6)(1/6)(1/6)\cdots = 0.$$

Another sequence where the limit is 2 is when $D_i = 1$ for i even and $D_i = 3$ for i odd. There are an uncountably infinite number of such sequences where the limit is not 3.5, or does not exist. The Strong Law of Large Numbers says that the probability of all of these bad sequences combined is always 0, which is very powerful!

CHAPTER 14: MEAN OF A RANDOM VARIABLE

LINEARITY OF EXPECTATION

Some operators have the property that they are *linear operators*. A linear operator takes as input a vector and returns either a vector or a scalar.

> **WHAT IS A VECTOR SPACE?**
> If you have not worked much before with vectors or scalars, here is the important part. A *vector space* consists of vectors and scalars. Two vectors can be added together to give another vector. A vector can also be scaled to give another vector by multiplying by a scalar. These operations have to obey certain rules of commutivity, associativity, and distribution.

> **D66** Say that $\mathcal{L}$ is a **linear operator** if for all vectors v and w, and scalars a and b,
> $$\mathcal{L}(av + bw) = a\mathcal{L}(v) + b\mathcal{L}(w).$$

Examples of linear operators include the following.

Derivatives. Here differentiable functions are the vectors and real numbers are the scalars. For f and g differentiable and a and b real numbers:

$$[af + bg]' = af' + bg'.$$

Integrals. Here integrable functions are the vectors and again real numbers are the scalars. For f and g integrable over A, and a and b real numbers:

$$\int_A af + bg \, dA = a\int_A f \, dA + b\int_A g \, dA.$$

Limits. Suppose a_n and b_n are sequences that both have a limit as n approaches ∞. Then for $c_1, c_2 \in \mathbb{R}$,

$$\lim_{n\to\infty} c_1 a_n + c_2 b_n = c_1 \lim_{n\to\infty} a_n + c_2 \lim_{n\to\infty} b_n.$$

But wait a minute! The Strong Law of Large Numbers says that the limit of the sample average equals the expectation with probability 1 when the expected value exists. So from the linearity of limits, the linearity of expectation follows.

> **F48** For integrable random variables as vectors, and real numbers as scalars, the expected value $\mathbb{E}$ is a linear operator.

Note that there is nothing about whether or not the random variables are dependent or independent. Linearity works in both cases! (Similarly, $[f + f]' = f' + f'$ even though the sum was of the same function.)

> **E48** Suppose X has expected value 3 and Y has expected value -1. What is $\mathbb{E}[6X + 4Y]$?
> **Answer.** By linearity of expectation, this is $6\mathbb{E}[X] + 4\mathbb{E}[Y]$, which is $(3)(6) + (-1)(4)$, or $\boxed{14}$.

> **E49** Let $U \sim \text{Unif}([0, 1])$, $X = 2\mathbb{I}(U < 0.3)$ and $Y = 3\mathbb{I}(U < 0.6)$. What is $\mathbb{E}[X + Y]$?
> **Answer.** Even though X and Y are not independent,
> $$\mathbb{E}[X + Y] = \mathbb{E}[X] + \mathbb{E}[Y]$$
> $$= 2\mathbb{E}[\mathbb{I}(U < 0.3)] + 3\mathbb{E}[\mathbb{I}(U < 0.6)]$$
> $$= (2)(0.3) + 3(0.6),$$
> which is $\boxed{2.400}$.

SYMMETRY

Some distributions are *symmetric*. When a random variable X is symmetric around a value m, the expected value of X is also m.
To be precise, first define what it means for a random variable to be symmetric around m.

> **D67** A random variable X is symmetric around m if $X - m$ and $-(X - m)$ have the same distribution.

> **E50** For $X \sim \text{Unif}(\{1, 2, 3, 4\})$, prove that X is symmetric around 2.5.
> **Answer.** Here $X - m$ is uniform over $\{-1.5, -0.5, 0.5, 1.5\}$, while $-(X - m)$ is uniform over $\{1.5, 0.5, -0.5, -1.5\}$, which is the same set!

In general, uniforms have the following symmetry.

> **F49** If $X \sim \text{Unif}(\{a, a + 1, \ldots, b\})$, then X is symmetric around $(a + b)/2$.

> **F50** If X is an integrable random variable symmetric around m, then $\mathbb{E}[X] = m$.

Proof. Note $\mathbb{E}[X - m] = \mathbb{E}[-(X - m)]$. So
$$0 = \mathbb{E}[X - m - [-(X - m)]]$$
$$= \mathbb{E}[2X - 2m]$$
$$= 2(\mathbb{E}(X) - m).$$

Solving for $\mathbb{E}[X]$ completes the proof. $\square$

Symmetric random variables with densities have symmetric densities.

> **F51** Let X be a random variable with density $f_X(x)$ symmetric around m. Then
> $$f_X(x - m) = f_X(m - x).$$

Proof. The density of $X - m$ is $f_X(x - m)$ and the density of $-(X - m) = m - X$ is $f_X(m - x)$. $\square$

WRITING SUMS AS INTEGRALS

Note that when $\mathbb{E}[X]$ exists for X discrete, the random variable is still called *integrable*, and not *summable*. This is because the sum can be written as an integral with respect to counting measure. Recall that

$$\sum_w w\mathbb{P}(W=w) = \int_w w f_W(w)\, d\#,$$

where # is counting measure. This integral formulation will be helpful when extending the idea of expectation to continuous random variables.

ENCOUNTERS

161. MEAN OF FINITE RANDOM VARIABLES

If $\mathbb{P}(W=1) = \mathbb{P}(W=2) = 0.13$, and $\mathbb{P}(W=4) = 0.74$, what is $\mathbb{E}[W]$?

162. ANOTHER MEAN OF A FINITE RANDOM VARIABLE

If $\mathbb{P}(R=-3) = 0.25$, $\mathbb{P}(R=0) = 0.4$, and $\mathbb{P}(R=1) = 0.35$, what is $\mathbb{E}[R]$?

163. MEAN OF A DISCRETE UNIFORM

Suppose $U \sim \mathsf{Unif}(\{0, 10, 100\})$. What is $\mathbb{E}[U]$?

164. ANOTHER MEAN OF A DISCRETE UNIFORM

Suppose A is uniform over $\{-1, 0, 1, 5\}$. What is $\mathbb{E}[A]$?

165. MEAN OF A DIE ROLL

Suppose $X \sim \mathsf{d}8$. What is $\mathbb{E}[\mathbb{I}(X \leq 3)]$?

166. MEAN OF A SUM

Suppose $U \sim \mathsf{Unif}([0,1])$, $X = \mathbb{I}(U \in [0.2, 0.3])$ and $Y = \mathbb{I}(U \in [0.25, 0.35])$. Find $\mathbb{E}[X + Y]$.

167. MEAN OF A DIFFERENCE

Suppose $\mathbb{E}[A] = 1.2$ and $\mathbb{E}[B] = 6.3$. What is $\mathbb{E}[A - B]$?

168. MEAN OF A FUNCTION OF A RANDOM VARIABLE

Let $W \sim \mathsf{Unif}(\{1, 2, 3\})$.
a. What is the density of $T = W^2$?
b. What is $\mathbb{E}[W]$?
c. What is $\mathbb{E}[T]$?
d. What is $\mathbb{E}[W + T]$?

169. SYMMETRY OF A DISCRETE UNIFORM

Suppose W is uniform over $\{-5, 0, 5\}$.
a. Show that W is symmetric around 0.
b. What is the expected value of W?

170. A SYMMETRIC FINITE RANDOM VARIABLE

Suppose $\mathbb{P}(R = 1) = \mathbb{P}(R = 3) = 0.4$, while $\mathbb{P}(R = 2) = 0.2$.
a. Show that R is symmetric around 2.
b. What is the expected value of R?

171. VERTIGON'S ARMY

The Dark Lord Vertigon was believed to have (with equal probability) a thousand, six thousand, or eight thousand soldiers in his army. What was the expected size of Vertigon's Army?

172. A PAPER MILL

A local paper mill expects with probability 40% to receive 1100 orders, with probability 15% to receive 800 orders, and with probability 45% to receive 600 orders. On average, how many orders do they expect to receive?

173. FOUR STORES

Pretty Polly's Pet Store has four locations. The first averages 200 customers a day, the second averages 232, the third 330, and the last 280. Altogether, what is the total average number of customers at all of the four stores per day?

174. STREAMLINING

Currently factory 1 produces an average of 10000 units per day, while factory 2 produces an average of 12000 units per day. A consultant believes that factory 1 can be improved by 20%, and factory 2 can be improved by 10%. If both statements are true, what would the average total output per day be?

175. THE SLLN IN ACTION

Suppose that W has mean 2 and $W_1, W_2, \ldots$ are iid W. What can be said about
$$\lim_{n \to \infty} \frac{W_1 + \cdots + W_n}{n}?$$

176. TO INFINITY!

Suppose X has mean 4.2 and $X_1, X_2, \ldots$ are iid X. What can you say about
$$\lim_{n \to \infty} \frac{X_1 + \cdots + X_n}{n}?$$

177. TWO TASKS

Suppose a certain task can be broken down into two parts. Part 1 requires T_1 amount of time and Part 2 requires T_2 amount of time. Part 2 cannot be started until Part 1 is complete. If $\mathbb{E}[T_1] = 4.2$ hours and $\mathbb{E}[T_2] = 1.3$ hours, what is the average time needed to complete both tasks?

178. A WEIGHTY MATTER

The King has received 40 ornaments from a crafter, but suspects that one ornament is a bit light on the gold. The weight of a proper ornament is known. The King's Vizier suggest the following plan. Break the ornaments into four groups on ten. Weigh each group until one is found that is below the target weight. Then go through each group one by one until the light ornament is found.

If the weighings are done by picking groups uniformly from unweighed groups until the light group is found, and then picking ornaments uniformly from unweighed ornaments in the light group until the light ornament is found, on average how many weighings are needed to locate the light ornament?

CHAPTER 14: MEAN OF A RANDOM VARIABLE

Chapter 15: Mean of a General Random Variable

The Wizard threw yet another fireball at the approaching monsters, then glanced backwards. Where were the reinforcements so desperately needed? The Wizard had modeled the time until the reinforcements came as exponentially distributed with rate parameter 0.3 per minute. With that model, what was the expected number of minutes until they arrived?

Expectations Using Densities

Last time an important fact was stated: the mean of an indicator function is just the probability that the event inside the indicator function is true.

So for a random variable X and measurable set A,

$$\mathbb{P}(X \in A) = \mathbb{E}[\mathbb{I}(X \in A)].$$

Of course, there is another way to find probabilities, use the density of X.

$$\mathbb{P}(X \in A) = \int_A f_X(x) \, d\mu = \int \mathbb{I}(x \in A) f_X(x) \, d\mu.$$

So that means

$$\mathbb{E}[\mathbb{I}(X \in A)] = \int \mathbb{I}(x \in A) f_X(x) \, d\mu.$$

Using linearity of expectations, this can be extended to weighted sums of indicator functions. So to find

$$\mathbb{E}[3\mathbb{I}(X \in A) + 2\mathbb{I}(X \in B)],$$

use

$$\int [3\mathbb{I}(x \in A) + 2\mathbb{I}(x \in B)] f_X(x) \, d\mu.$$

In general, if $h(x)$ is any weighted sum of indicator functions:

$$h(x) = a_1 \mathbb{I}(x \in A_1) + \cdots + a_n \mathbb{I}(x \in A_n),$$

then

$$\mathbb{E}[h(X)] = \int h(x) f_X(x) \, dx.$$

That tells us how to find the expected value of a function of a random variable with a density when the function is the weighted sum of indicator functions.

Even if the target function is not exactly a weighted sum of indicator functions, a weighted sum of indicator functions might be good enough to approximate the function.

For instance, consider the following four interval partition of $[0, 1]$.

$$I_0 = [0, 1/4],$$
$$I_1 = (1/4, 1/2],$$
$$I_2 = (1/2, 3/4],$$
$$I_3 = (3/4, 1].$$

Then
$$\sum_{i=0}^{3}\left(\frac{i}{4}\right)^2 \mathbb{I}(x \in I_i)$$
is an approximation to $y = x^2$ over the interval $[0,1]$.

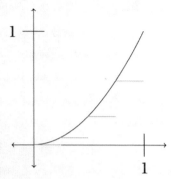

By using more indicator functions, the approximation gets better and better. This leads to the following way to calculate expected value for random variables with densities.
$$\mathbb{E}[g(X)] = \int g(x) f_X(x)\, d\mu.$$

In general, this result is one of the most important in probability theory.

T5 Suppose random variable X has density f_X with respect to measure μ. Then for a real-valued function g,
$$\mathbb{E}[g(X)] = \int g(x) f_X(x)\, d\mu(x),$$
if the integral exists.

Recall that our two most common measures are *counting measure* for discrete random variables, and *Lebesgue measure* for continuous random variables. For Lebesgue measure
$$\mathbb{E}[g(X)] = \int g(x) f_X(x)\, dx,$$
and for counting measure,
$$\mathbb{E}[g(X)] = \int g(x) f_X(x)\, d\# = \sum_x g(x) f_X(x).$$

This is summarized in the following result.

F52 Let X be any continuous random variable with density $f_X(x) = \mathbb{P}(X \in dx)/dx$ with respect to Lebesgue measure. Then
$$\mathbb{E}[g(X)] = \int g(x) f_X(x)\, dx.$$

For X a discrete random variable with density $f_X(i) = \mathbb{P}(X = i)$ with respect to counting measure
$$\mathbb{E}[g(X)] = \sum g(i) f_X(i).$$

This result is sometimes called the *Law of the Unconscious Statistician* since every instance of the random variable X is replaced by the index variables in the integral or sum. Some examples!

E51 Set up the integral for $\mathbb{E}[\exp(T)]$, where $T \sim \mathsf{Unif}([0, 10])$.
Answer. Note T is a continuous random variable with density
$$f_T(t) = \frac{1}{10}\mathbb{I}(t \in [0,10]).$$
Hence
$$\mathbb{E}[\exp(T)] = \int \exp(t)(1/10)\mathbb{I}(t \in [0,10])\, dt$$
which is
$$\boxed{\int_0^{10} \exp(t)/10\, dt.}$$

E52 Set up the integral for $\mathbb{E}[W^2]$, where $W \sim \mathsf{Unif}(\{1,\ldots,5\})$.
Answer. Note $\mathbb{P}(W = i) = 1/5$ for $i \in \{1, 2, \ldots, 5\}$. Hence this is
$$\boxed{\sum_{i=1}^{5} i^2 (1/5).}$$

E53 Set up the integral for $\mathbb{E}[\sqrt{T}]$, where $T \sim$

Exp(1).
Answer. The density of T is $f_T(t) = \exp(-t)\mathbb{I}(t \geq 0)$, so the integral is

$$\mathbb{E}[\sqrt{T}] = \int \sqrt{t}\exp(-t)\mathbb{I}(t \geq 0)\, dt$$

$$= \boxed{\int_0^\infty \sqrt{t}\exp(-t)\, dt.}$$

SOLVING THE STORY

In the case of the story, the goal was to find $\mathbb{E}[T]$ where $T \sim \text{Exp}(0.3 \text{ per minute})$. The density is

$$f_T(t) = 0.3\exp(-0.3t)\mathbb{I}(t \geq 0),$$

and $g(t) = t$, so the expected value is

$$\mathbb{E}[T] = \int t \cdot 0.3\exp(-0.3t)\mathbb{I}(t \geq 0)\, dt$$

$$= \int_0^\infty t \cdot 0.3\exp(-0.3t)\, dt$$

$$= \int_0^\infty t[-\exp(-0.3t)]'\, dt$$

$$= -t\exp(-0.3t)\big|_0^\infty - \int_0^\infty -[t]'\exp(-0.3t)\, dt$$

$$= -0.3^{-1}\exp(-0.3t)\big|_0^\infty$$

$$= 0.3^{-1}$$

which is about $\boxed{3.333 \text{ minutes}}$.

SYMMETRY

Earlier it was noted that for integrable random variables that are symmetric about m, the mean must also be m as well. This works equally well for continuous as well as discrete random variables.

> **F53** Suppose X has density $f_X(x)$ with respect to Lebesgue measure, where $f_X(x-m) = f_X(m-x)$ for some constant m. Then if X is integrable, then $\mathbb{E}[X] = m$.

Both discrete uniform random variables over $\{a, a+1, \ldots, b-1, b\}$ and continuous uniform random variables over $[a,b]$ are symmetric about $(a+b)/2$. Moreover, each has a finite expected value because they are bounded below by a and above by b. Together, this gives the following result.

> **F54** Suppose $X \sim \text{Unif}(A)$, where $A = \{a, a+1, \ldots, b\}$ or $A = [a,b]$. Then
> $$\mathbb{E}[X] = \frac{a+b}{2}.$$

MONTE CARLO METHODS

The term *Monte Carlo method* (MCM) refers to any algorithm that draws from random variables while running.

One type of MCM can use the random draws to estimate sums or integrals. For instance, suppose the goal is to estimate the integral

$$I = \int_0^\infty t^{3/2}\exp(-t)\, dt.$$

Note that $I = \mathbb{E}[T^{3/2}]$, where T is an exponential random variable with mean 1. So use the Strong Law of Large Numbers!

Draw n iid copies of T where n is a large number, then take the sample average of the results. In R, this can be accomplished with the following code.

```
n <- 10^6
T <- rexp(n, 1)
print(mean(T^(3/2)))
```

The result is close to the true answer of 1.32934.

ENCOUNTERS

179. MEAN OF AN EXPONENTIAL

For $W \sim \text{Exp}(-2)$, set up the following integrals.

a. $\mathbb{E}[W]$.
b. $\mathbb{E}[W^2]$.
c. $\mathbb{E}[\mathbb{I}(W < 3)]$

180. Mean of a uniform

For $U \sim \mathsf{Unif}([-5, 5])$, set up the following integrals.
a. $\mathbb{E}[U]$.
b. $\mathbb{E}[U^2]$.
c. $\mathbb{E}[\mathbb{I}(U < 3)]$

181. Mean of functions of a continuous uniform

For a random variable $T \sim \mathsf{Exp}(\lambda)$, so $f_T(t) = \lambda \exp(-\lambda t)\mathbb{I}(t \geq 0)$, find $\mathbb{E}[T]$.

182. Mean of a general exponential

Consider a random variable $T \sim \mathsf{Exp}(\lambda)$ with density $f_T(t) = \lambda \exp(-\lambda t)\mathbb{I}(t \geq 0)$. Find $\mathbb{E}[T^2]$.

183. Monte Carlo with uniforms

Using the function `runif` that generates $U \sim \mathsf{Unif}([0, 1])$, write R code to estimate
$$\int_0^1 \sqrt{u} \, du$$
using 10^6 samples.

184. Monte Carlo with exponentials

Using 'rexp' that generates random variables with an exponential distribution, write R code to estimate
$$\int_0^\infty \sin(x) \exp(-0.5x) \, dx.$$

185. Polynomials of continuous uniforms

Let $U \sim \mathsf{Unif}([0, 1])$. Find $\mathbb{E}[(1 - U)(1 + U)]$.

186. Polynomials of discrete uniforms

Let $W \sim \mathsf{Unif}(\{1, 2, 3\})$. Find $\mathbb{E}[(1 - W)(1 + W)]$.

187. A nonintegrable density

Suppose X has density $f_X(x) = (4/\tau)/(1 + x^2)$. Show that X is not integrable.

188. A nonintegrable random variable

Show that if $U \sim \mathsf{Unif}((0, 1])$, then U^{-1} is not integrable.

189. Deriving formulas

Suppose $\mathbb{E}[X] = \mu$. Prove that
$$\mathbb{E}[(X - \mu)^2] = \mathbb{E}[X^2] - \mu^2.$$

190. A cubic formula

For a random variable X with X^3, X^2, and X integrable random variables, let $\mu = \mathbb{E}[X]$. Then write
$$\mathbb{E}[(X - \mu)^3]$$
in terms of $\mathbb{E}[X^3]$, $\mathbb{E}[X^2]$, and μ.

Chapter 16: Conditional Expectation

HE FIGHTER QUICKLY SIZED UP the approaching monsters. Based on the size of the cave, the Fighter figured that the monsters were composed of from 1 to 3 groups, and each possibility was equally likely. Each group of monsters was equally likely to contain from 10 to 20 individuals. Overall, the Fighter wondered, what is the average number of monsters that approached?

Knowing about a random variable

In order to solve the Fighter's problem, it is necessary to understand how *conditional expectation* works. Recall that for conditional probability, the goal was to find the probability that an event s was true given that it was known that r was true.

Expectation can be thought of as an extension of probability since $\mathbb{P}(s) = \mathbb{E}[\mathbb{I}(s)]$. So naturally conditional expectation is a bit more complicated than conditional probability.

The idea is as follows. Consider random variables X and Y which are both integrable. Then knowing nothing about Y, X has a particular average. But if the value of Y is known instead of random, it might be possible to say more about the value of X.

For instance, in the story, let G be the number of groups of monsters. Then $G \sim \mathsf{Unif}(\{1,2,3\})$. Let N_1 be the number of monsters in the first group, N_2 the number in the second group (if there is a second group), and N_3 be the number in the third group (if there is a third group). Each N_i is uniform over $\{10, 11, \ldots, 20\}$.

Then let T be the total number of monsters. Then the description of T depends on the value of G.

$$T = N_1 \qquad \text{if } G = 1$$
$$T = N_1 + N_2 \qquad \text{if } G = 2$$
$$T = N_1 + N_2 + N_3 \qquad \text{if } G = 3$$

Or this can be written in a single line as:
$$T = \sum_{i=1}^{G} N_i.$$

If the value of G was known, then it would be possible to find the mean of T. For instance,

$\mathbb{E}[T \mid G = 1] = \mathbb{E}[N_1] = 15$
$\mathbb{E}[T \mid G = 2] = \mathbb{E}[N_1 + N_2] = \mathbb{E}[N_1] + \mathbb{E}[N_2] = 30$
$\mathbb{E}[T \mid G = 3] = \mathbb{E}[N_1] + \mathbb{E}[N_2] + \mathbb{E}[N_3] = 45.$

Another way to say this is that
$$\mathbb{E}[T \mid G] = 15G.$$

Plugging 1 or 2 or 3 in for G gives the correct answer in the right hand side.

This should make sense, if G is known, then the mean of T is just the value 15 summed G times, which is $15G$.

The value of $\mathbb{E}[T \mid G]$ is itself a random variable, in fact it is $h(G)$, where $h(g) = 15g$ is a simple function.

So what is the next step in solving the story? That would be the Fundamental Theorem of Probability.

The Fundamental Theorem of Probability

Probability is about information, and conditioning is the way that information is presented in probability problems. So it should not be a surprise that the most important theorem about conditioning should be the Fundamental Theorem of Probability.

The FTP works as follows. Given knowledge of random variable Y, random variable X has a mean that is denoted $\mathbb{E}[X \mid Y]$. That is itself a random variable that is a function of Y.

Now suppose that Y was not known, but still a random variable. Suppose the value of $\mathbb{E}[X \mid Y]$ is averaged over all possible values that Y could take on. The FTP says that the result, will be the average over X.

In short, the average over averages of X with partial information Y, is the total average of X. This is the Fundamental Theorem of Probability.

T6 The Fundamental Theorem of Probability
Given an integrable random variable X,
$$\mathbb{E}[\mathbb{E}[X \mid Y]] = \mathbb{E}[X].$$

Solving the Story

Back to the Story. The total number of monsters given the number of groups is
$$\mathbb{E}[T \mid G] = 15G.$$
But the goal is to find $\mathbb{E}[T]$. The Fundamental Theorem of Probability says that the conditioning can be undone by taking the mean of both sides of the equation again.
$$\mathbb{E}[\mathbb{E}[T \mid G]] = \mathbb{E}[15G] = 15\mathbb{E}[G].$$

Since $G \sim \text{Unif}(\{1,2,3\})$, $\mathbb{E}[G] = (1+3)/2 = 2$. Hence
$$\boxed{\mathbb{E}[T] = 30.}$$

This seems reasonable: there are an average of two groups of monsters and each group of monsters contains on average 15 members. So the overall average number of monsters is $2 \cdot 15 = 30$. That is what makes the Fundamental Theorem of Probability so compelling; once you realize what it is saying, it makes sense in a very deep way.

Properties of Conditional Expectation

There is another way to view conditional expectation. The idea is that when a random variable Y is part of the conditioning, it behaves more like a constant than a random variable. Then the rules for constants apply to the Y random variable. This gives some properties of conditional expectations that assist in evaluation. These extend the rules already in place for expected value to conditional expected value.

F55 Properties of conditional expectation.

1. Linearity. For a and b constants, X, Y, and Z random variables,
$$\mathbb{E}[aX + bY \mid Z] = a\mathbb{E}[X \mid Z] + b\mathbb{E}[Y \mid Z].$$

2. Independence. If X and Y are independent random variables then
$$\mathbb{E}[X \mid Y] = \mathbb{E}[X].$$

3. Multiplication. For any deterministic function f,
$$\mathbb{E}[f(Y)X \mid Y] = f(Y)\mathbb{E}[X \mid Y].$$

A useful special case of the last result is that if $f(Y) = Y$ and $X = 1$ with probability 1, then
$$\mathbb{E}[Y \mid Y] = Y.$$

E54 Suppose $\mathbb{E}[X \mid Y] = 5Y^2$ and $\mathbb{E}[W \mid Y] = 3Y^2$. What is $\mathbb{E}[X - Y + W \mid Y]$?
Answer. By linearity, the answer a can be written as
$$a = \mathbb{E}[X - Y + W \mid Y]$$
$$= \mathbb{E}[X \mid Y] - \mathbb{E}[Y \mid Y] + \mathbb{E}[W \mid Y].$$

Using the values given in the problem together with the fact that $\mathbb{E}[Y \mid Y] = Y$ gives
$$\boxed{\mathbb{E}[X - Y + W \mid Y] = 8Y^2 - Y.}$$

EXPECTATION AND PROBABILITY TREES

In the FTP, $\mathbb{E}[X] = \mathbb{E}[\mathbb{E}[X \mid Y]]$. In the right hand side, there is an inner mean $\mathbb{E}[X \mid Y]$ and an outer mean $\mathbb{E}[\cdots]$.

Earlier, the inner mean was evaluated first and then the outer mean. For some problems, it helps to evaluate the outer mean first, and then the inner mean.

E55 The price of a tech stock is highly dependent on the larger economy. A particular stock is modelled as follows.

When the economy is good (which happens with probability 20%), the stock will grow 30%. when the economy is medium (45%), the stock will grow 10%. when the economy is poor (35%), the stock will shrink by 10%.

What is the average the stock grows?
Answer. For a random variable W that takes on discrete values,
$$\mathbb{E}[W] = \sum_i i \mathbb{P}(W = i).$$

Create a random variable X for the economy where
$$(X = 1) = \text{the economy is good}$$
$$(X = 2) = \text{the economy is medium}$$
$$(X = 3) = \text{the economy is poor}$$

and let S be the fractional growth in the stock. Then the goal is to find
$$\mathbb{E}[S] = \mathbb{E}[\mathbb{E}[S \mid X]].$$

The key is to see that $\mathbb{E}[S \mid X]$ takes on three values, it will be one of
$$\mathbb{E}[S \mid X = 1] = 30\%$$
$$\mathbb{E}[S \mid X = 2] = 10\%$$
$$\mathbb{E}[S \mid X = 3] = -10\%$$

So that means that our answer a is
$$a = \mathbb{E}[\mathbb{E}[S \mid X]]$$
$$= \sum_{i=1}^{3} \mathbb{E}[S \mid X = i] \mathbb{P}(X = i)$$
$$= (0.3)(0.2) + (0.45)(0.1) + (0.35)(-0.1),$$
which is $\boxed{7\%}$.

In order to keep straight what is happening in this kind of argument, a graphical form called *expectation trees* can be used.

To find $\mathbb{E}[\mathbb{E}[X \mid Y]]$, a branch is created for each value that Y can take on. The branch is labeled with the probability that Y takes on that value. At the end of the branch, the value of X given that Y value is written.

In the last example, the expectation tree looks as follows.

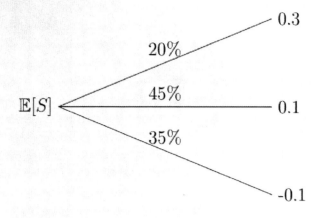

The *value* of the expectation tree is the sum of the product of the edge labels with the value at the end of the branch. In the tree above, this gives

$$\mathbb{E}[S] = (0.2)(0.3) + (0.45)(0.1) + (0.35)(-0.1)$$
$$= 0.07$$

as above.

Remember that because of the relationship $\mathbb{E}[\mathbb{I}(s)] = \mathbb{P}(s)$, expectations are generalizations of probabilities. When using an expectation tree to compute probabilities, this is called a *probability tree*. The algebraic version of this is called the *Law of Total Probability*. Recall that this says that if $s_1, s_2, \ldots$ are disjoint events such that exactly one must be true, then for any statement r,

$$\mathbb{P}(r) = \sum_{i=1}^{\infty} \mathbb{P}(r \mid s_i)\mathbb{P}(s_i).$$

The expectation of a Geometric random variable

Recall that $B_1, B_2, \ldots$ is a *Bernoulli process* of rate p if the variables are iid Bernoulli random variables with mean p.

For such a process, let

$$G = \inf\{i : B_i = 1\}.$$

Then G is a geometric random variable, and write $G \sim \mathsf{Geo}(p)$.

The FTP can be used to find the mean of a geometric random variable.

F56 The mean of $G \sim \mathsf{Geo}(p)$ is $1/p$.

Proof. Condition on the first Bernoulli B_1. Then the probility that $B_1 = 1$ is p, and the probability that $B_1 = 0$ is $1 - p$. Hence

$$\mathbb{E}[G] = \mathbb{E}[\mathbb{E}[G \mid B_1]]$$
$$= \mathbb{E}[G \mid B_1 = 1]p + \mathbb{E}[G \mid B_1 = 0](1-p).$$

If $B_1 = 1$, then $G = 1$ because G is counting the number of draws until a 1 appears.

If $B_1 = 0$, then the first flip is wasted, so it is neccessary to wait another G steps to get the first one. In other words, $[G \mid B_1 = 0] \sim 1 + G$.

Hence

$$\mathbb{E}[G] = (1)(p) + (\mathbb{E}[G] + 1)(1-p).$$

Solving for $\mathbb{E}[G]$ gives $\mathbb{E}[G] = 1/p$. □

Connection to conditional probability

Expectation generalizes probability through the relationship

$$\mathbb{P}(s) = \mathbb{E}[\mathbb{I}(s)]$$

for an event s. Similarly, conditional expectation generalizes conditional probability. To see why this is true, suppose $X = \mathbb{I}(s)$ and $Y = \mathbb{I}(r)$. Then

$$\mathbb{E}[XY] = \mathbb{E}[\mathbb{E}[XY \mid Y]]$$

There are only two possible values for Y, 1, and 0. For $Y = 1$,

$$\mathbb{E}[XY \mid Y = 1] = \mathbb{E}[X \cdot 1 \mid Y = 1]$$
$$= \mathbb{E}[X \mid Y = 1]$$

For $Y = 0$,
$$\mathbb{E}[XY \mid Y = 0] = \mathbb{E}[X \cdot 0] = 0.$$

That means
$$\mathbb{E}[XY] = \mathbb{E}[\mathbb{E}[XY \mid Y]] \\ = \mathbb{E}[X \mid Y = 1]\mathbb{P}(Y = 1) + (0)\mathbb{P}(Y = 1).$$

Here
$$XY = \mathbb{I}(s)\mathbb{I}(r) = \mathbb{I}(sr).$$

Conditioning on $Y = 1$ is the same as conditioning on r being true. So
$$\mathbb{E}[X \mid Y = 1] = \mathbb{P}(s \mid r).$$

Hence
$$\mathbb{E}[\mathbb{I}(sr)] = \mathbb{E}[\mathbb{I}(s) \mid r]\mathbb{P}(r),$$

and since the mean of an indicator function is the probability that the thing inside the indicator function occurs,
$$\mathbb{P}(sr) = \mathbb{P}(s \mid r)\mathbb{P}(r).$$

Therefore, conditional probability is a special case of conditional expectation in the same way as regular probability is a special case of expectation.

ENCOUNTERS

191. RAISING SALES

A marketing firm believes that an ad campaign has a 30% chance of raising sales exactly 100%, a 50% chance of raising sales exactly 20%, and a 20% chance of having no effect on sales.
a. Calculate the expected raise in sales.
b. Now suppose that instead of exactly raising sales 100%, 20% and 0%, there is a 30% chance of raising sales a random amount that has average value 100%, a 50% chance of raising sales a random amount that has average 20%, and a 20% chance of raising sales a random amount that has average 0%. Now calculate the expected raise in sales.

192. FIRE!

Forest fire season in an area sees on average 15 000 acres burn if rainfall is high, 30 000 acres burn if rainfall is medium, and 75 000 acres if rainfall is low. If the probability of high, medium, and low rainfall are 0.34, 0.47, and 0.19 respectively, what is the average number of acres burned in a season?

193. FOOD FOR THOUGHT

The weather for a region is unusually rainy with probability 40%, in which case the chance of a farmer obtaining a full crop is 60%. It is normal precipitation with probability 40%, in which case there is a 90% chance of obtaining a full crop. Finally, there is a 20% chance of lower precipitation, which gives a 40% chance of a full crop.

a. Draw a probability tree for the probability of getting a full crop.
b. Calculate the overall probability of getting a full crop.

194. THE SELL OUT

A manufacturer will sell out of inventory with probability 90% if demand is high for a product, and will sell out with probability 50% if demand is low.

a. Draw a probability tree to evaluate the probability of selling out of inventory.
a. Find the probability of selling out of inventory.

195. RANDOM NUMBERS OF DICE

Let $X_1, X_2, \ldots$ be iid Unif($\{1, 2, 3\}$).
a. What is $\mathbb{E}[X_i]$?
b. What is $\mathbb{E}[X_1 + X_2 + X_3 + X_4]$?

c. If $G \sim \text{Geo}(0.3)$, what is

$$\mathbb{E}\left[\sum_{i=1}^{G} X_i\right].$$

196. Random number of exponentials

Suppose $\mathbb{P}(Y = 1) = 0.2$, $\mathbb{P}(Y = 3) = 0.2$, $\mathbb{P}(Y = 7) = 0.8$. Let $X_1, \ldots, X_8$ be iid $\text{Exp}(2.1)$. Find

$$\mathbb{E}\left[\sum_{i=1}^{Y} X_i\right].$$

197. The Bayesian, Part I

A Bayesian statistician models a parameter θ as having an $\text{Exp}(2)$ distribution. The data X given θ is uniform over $[0, \theta]$.
a. What is $\mathbb{E}[X|\theta]$?
b. What is $\mathbb{E}[X]$?

198. The Bayesian, Part II

A statistical model has $X \sim \text{Exp}(\lambda)$, where $\lambda \sim \text{Unif}([3, 4])$. What is $\mathbb{E}[X]$?

199. Using the FTP

The FTP applies to multiple conditions as well. So for instance $\mathbb{E}[X] = \mathbb{E}[\mathbb{E}[X|Y, Z]]$.
  Suppose $Z \sim \text{Unif}([1, 2])$, $[Y|Z] \sim \text{Exp}(Z)$ and $[X|Y, Z] \sim \text{Unif}(Z, Z + Y)$. Find $\mathbb{E}[X]$.

200. More FTP

Suppose $X \sim \text{Unif}([10, 20])$, $[Y|X] \sim \text{Unif}[0, X]$, and $[T|Y] \sim \text{Exp}(Y)$.
  a. What is $\mathbb{E}[Y]$?
  b. What is $\mathbb{E}[T]$?

Chapter 17: Joint Densities of Random Variables

A Ranger was waiting for two comrades to arrive. Bored, the Ranger began to model their arrival times. The first step was to give the arrival times names: T_1 for the arrival time of the first comrade and T_2 for the arrival time of the second comrade.

Furthermore, the Ranger modeled T_1 and T_2 using a joint density,

$$f_{(T_1,T_2)}(t_1, t_2) = (3/2)\exp(-3t_1 - t_2/2)\mathbb{I}(t_1, t_2 \geq 0).$$

What is $\mathbb{P}((T_1, T_2) \in [0, 0.6] \times [0, 1.5])$?

Univariate Random Variables

A random variable $X \in \mathbb{R}$ in one dimension is called *univariate*, where uni here is the Latin prefex for one. A helpful mnemonic to remember that a univariate random variable has only a single value, is that unicorns have only a single horn.

If a random variable lives in a higher dimension, for instance (T_1, T_2) as in the Story for today, the random variable is called *multivariate*.

In this section, it will be shown how the notion of densities for univariate random variables can be extended to the multivariate case. This means some higher dimensional sums and integrals are coming, so review that Multivariable Calculus course you took and get ready for higher dimensions!

Densities in Higher Dimensions

Recall that *differential notation* $\mathbb{P}(X \in da)$ means the probability that the random variable X is in a small differential element around the value a. So far only random variables in one dimension have been considered. In this section these differential elements will be in two or more dimensions.

Write

$$\mathbb{P}(X \in da) = f_X(a)\, da$$

to mean that the random variable X has *density* f_X. This holds even when X is a multidimensional variable, for instance it could be that $X = (X_1, X_2, X_3)$ and is three dimensional. In which case, you could write out the dimensions explicitly, using

$$\mathbb{P}(X_1 \in dx_1, X_2 \in dx_2, X_3 \in dx_3)$$

equals to

$$f_{(X_1, X_2, X_3)}(x_1, x_2, x_3)\, dx_1\, dx_2\, dx_3.$$

Just like in one dimension, the function $f_{(X_1, X_2, X_3)}$ on the right hand side is called the *density* of (X_1, X_2, X_3).

F57 Say that a random vector $(X_1, X_2, \ldots, X_n)$ has a **density** with respect to product measure $\mu_1 \times \cdots \times \mu_n$ if

$$\mathbb{P}(X_1 \in d\mu_1, \ldots, X_n \in d\mu_n)$$

is equal to

$$f_{(X_1,\ldots,X_n)}(x_1,\ldots,x_n)\, d\mu_1 \times \cdots \times d\mu_n.$$

INTEGRATION IN HIGHER DIMENSIONS
This differential notation means that if the goal is to find $\mathbb{P}((X_1,\ldots,X_n) \in A)$, just integrate the product measure over the region. So

$$\mathbb{P}((X_1,\ldots,X_n) \in A)$$

equals

$$\int_{(x_1,\ldots,x_n) \in A} f_{(X_1,\ldots,X_n)}(x_1,\ldots,x_n)\, d\mu_1 \times \cdots \times d\mu_n.$$

If the measure is counting measure, this integral just becomes a sum as in one dimension.

E56 Suppose $(A,B) \in \Omega$, where $\Omega = \{1,2\} \times \{1,2,3\}$. Then (A,B) has joint density

$$f_{(A,B)}(a,b) = \frac{(a+2b)}{33}\mathbb{I}((a,b) \in \Omega).$$

What is $\mathbb{P}(B \leq 2)$?
Answer. Sum the probabilities of elements of Ω with $b \leq 2$. These are the points $(1,1)$, $(2,1)$, $(1,2)$, $(2,2)$, which gives

$$\mathbb{P}(B \leq 2) = \sum_{(a,b) \in \Omega : b \leq 2} f_{(A,B)}(a,b)$$

The four terms in the sum are

$$f_{(A,B)}(1,1) + f_{(A,B)}(2,1) + f_{(A,B)}(1,2) + f_{(A,B)}(2,2),$$

which means

$$\mathbb{P}(B \leq 2) = \frac{3+4+5+6}{33} = \frac{18}{33} = \frac{6}{11},$$

so about $\boxed{54.54\%}$.

SOLVING THE STORY

In the Story, the density $f_{(T_1,T_2)}(t_1,t_2)$ was given as

$$(3/2)\exp(-3t_1 - t_2/2)\mathbb{I}(t_1, t_2 \geq 0).$$

So to find $\mathbb{P}(T_1 \in [0,0.6], T_2 \in [0,1.5])$ this joint density is integrated over this region. Because this is a density with respect to Lebesgue measure, the integral is over a subset of $\mathbb{R}^2$.

The probability

$$\mathbb{P}(T_1 \in [0,0.6], T_2 \in [0,1.5])$$

is equal to the integral

$$\int_{(t_1,t_2) \in A} (3/2)\exp(-3t_1 - t_2/2)\mathbb{I}(t_1, t_2 \geq 0)\, d\mathbb{R}^2,$$

where $A = [0,0.6] \times [0,1.5]$.

Integrals in one dimension are easier to do than integrals in multiple dimensions. A theorem named *Tonelli's Theorem* tells us that whenever the integrand is greater than or equal to 0 (as they always are with densities), you can use *iterated integrals* to find the integral. That means the problem of

$$p = \mathbb{P}(T_1 \in [0,0.6], T_2 \in [0,1.5])$$

can be written as an iterated integral

$$p = \int_{t_1=0}^{0.6} \int_{t_2=0}^{1.5} (3/2)\exp(-3t_1 - t_2/2) dt_2\, dt_1.$$

The total dimension must be the same for the original and the iterated integrals, That is, if you started with one two dimensional integral, you will end up with two one dimensional integrals in the iterated version.

To solve an iterated integral, work

from the inside out.

$$p = \int_{t_1=0}^{0.6} (3/2)(-2)\exp(-3t_1 - t_2/2)|_{t_2=0}^{1.5}\, dt_1$$

$$= \int_{t_1=0}^{0.6} -3[\exp(-3t_1 - 0.75) - \exp(-3t_1)]\, dt_1$$

$$= \int_{t_1=0}^{0.6} 3\exp(-3t_1)[1 - \exp(-0.75)]\, dt_1$$

$$= -\exp(-3t_1)[1 - \exp(-0.75)]|_{t_1=0}^{0.6}$$

$$= [1 - \exp(-1.8)][1 - \exp(-0.75)]$$

which is about $\boxed{0.4404}$.

MARGINAL DENSITIES

D68 Consider a random vector $(X_1, \ldots, X_n)$ where $n \geq 2$. Then the density of a particular X_i is called a **marginal density**.

To calculate these marginal densities, use the fact that $\mathbb{P}(X_1 \in dx_1)$ can be written as

$$\mathbb{P}(X_1 \in dx_1, X_2 \in \mathbb{R}, X_3 \in \mathbb{R}, \ldots, X_n \in \mathbb{R}).$$

This probability can be found by *integrating out* the variables that we do not want to be in the density. To find the density of X_1, integrate the joint density

$$f_{(X_1,\ldots,X_n)}(x_1, \ldots, x_n)$$

with respect to x_2, x_3, and so on up to x_n, all from negative infinity to positive infinity.

A useful notation for a vector v with component i removed is v_{-i}. For instance, if $v = (1, 2, 3, 4)$, then $v_{-3} = (1, 2, 4)$. That notation makes the following fact easier to see.

F58 Let $X = (X_1, \ldots, X_n)$ and $x = (x_1, \ldots, x_n)$. Given the density $f_X(x)$ of X, the density of X_{-i} at x_{-i} is

$$f_{X_{-i}}(x_{-i}) = \int_{x_i \in \mathbb{R}} f_X(x)\, d\mu_i.$$

If you want the remove all but one random variable in the vector, integrate out every random variable except the one that you wish to keep.

E57 Suppose as before that $(A, B) \in \Omega$, where $\Omega = \{1, 2\} \times \{1, 2, 3\}$ has density

$$f_{(A,B)}(a, b) = \frac{(a + 2b)}{33}\mathbb{I}((a, b) \in \Omega).$$

a. What is the density of A?
b. What is the density of B?

Answer. a. To get the density of A, we *integrate out* B. Because we are working on discrete variables, integration becomes summation:

$$f_A(a) = \sum_{b=1}^{3} f_{(A,B)}(a, b)$$

$$= \left[\frac{a+2}{33} + \frac{a+4}{33} + \frac{a+6}{33}\right]\mathbb{I}(a \in \{1, 2\}),$$

which simplifies to

$$\boxed{f_A(a) = \left[\frac{a+4}{11}\right]\mathbb{I}(a \in \{1, 2\}).}$$

b. To get the density of B we *integrate out* A. That gives

$$f_B(b) = \sum_{a=1}^{2} f_{(A,B)}(a, b)$$

$$= \left[\frac{1+2b}{33} + \frac{2+2b}{33}\right]\mathbb{I}(b \in \{1, 2, 3\}),$$

which reduces to

$$\boxed{f_B(b) = \frac{3+4b}{33}\mathbb{I}(b \in \{1, 2, 3\}).}$$

E58 Let $X = (X_1, X_2, X_3)$ have joint density $f_X(x)$, where $x = (x_1, x_2, x_3)$, of

$$(1/48)[x_1 + 2x_2 + 3x_3]\mathbb{I}(x_1, x_2, x_3 \in [0, 2])$$

What is the density of X_2?
Answer. Let $A = [0, 2] \times [0, 2] \times [0, 2]$, and

$$g(x) = \frac{1}{48}[x_1 + 2x_2 + 3x_3]$$

so

$$f_X(x) = g(x)\mathbb{I}(x \in A).$$

The indicator will change the limits of integration, and the $g(x)$ is the part that varies with x.

To find the density of X_2, it is necessary to integrate out both X_1 and X_3. This can be done as follows:

$$f_{X_2}(x_2) = \int_{x_1 \in \mathbb{R}} \int_{x_3 \in \mathbb{R}} g(x)\mathbb{I}(x \in A) \, dx_3 \, dx_1.$$

Use the indicator function to change the limits, so

$$f_{X_2}(x_2) = \mathbb{I}(x_2 \in [0, 2]) \int_{x_1=0}^{2} \int_{x_3=0}^{2} g(x) \, dx_3 \, dx_1$$

Do the inside integral first to get

$$I_3 = \int_{x_3=0}^{2} g(x) \, dx_3$$
$$= \frac{1}{48}[x_1 x_3 + 2x_2 x_3 + (3/2)x_3^2]\Big|_{x_3=0}^{2}$$
$$= \frac{1}{24}[x_1 + 2x_2 + 3].$$

Then integrate out x_1 to get

$$I_{1,3} = \int_{x_1=0}^{2} I_3 \, dx_1$$
$$= \int_{x_1=0}^{2} \frac{1}{24}[x_1 + 2x_2 + 3] \, dx_1$$
$$= \frac{1}{24}[\frac{1}{2}x_1^2 + 2x_2 x_1 + 3x_1]\Big|_{x_1=0}^{2}$$
$$= \frac{1}{24}[2 + 4x_2 + 6]\mathbb{I}(x_2 \in [0, 2]),$$

which means

$$\boxed{f_{X_2}(x_2) = \frac{2 + x_2}{6}\mathbb{I}(x_2 \in [0, 2]).}$$

SHOWING INDEPENDENCE USING DENSITIES

Suppose that X has density $f_X(x)$ with respect to μ_1, Y has density $f_Y(y)$ with respect to μ_2, and their joint density is $f_{(X,Y)}(x, y) = f_X(x)f_Y(y)$ with respect to $\mu_1 \times \mu_2$. Then consider sets A (measurable with respect to X) and B (measurable with respect to Y).

Then

$$p = \mathbb{P}(X \in A, Y \in B)$$
$$= \int_{(x,y) \in A \times B} f_{(X,Y)}(x, y) \, d[\mu_1 \times \mu_2]$$
$$= \int_{(x,y) \in A \times B} f_X(x)f_Y(y) \, d[\mu_1 \times \mu_2]$$
$$= \int_{x \in A} \int_{y \in B} f_X(x)f_Y(y) \, d\mu_2 \, d\mu_1$$
$$= \int_{x \in A} f_X(x) \int_{y \in B} f_Y(y) \, d\mu_2 \, d\mu_1$$
$$= \int_{x \in A} f_X(x)\mathbb{P}(Y \in B) \, d\mu_2$$
$$= \mathbb{P}(Y \in B) \int_{x \in A} f_X(x) \, d\mu_1$$
$$= \mathbb{P}(Y \in B)\mathbb{P}(X \in A).$$

This proves the following fact.

F59 Suppose that X has density $f_X(x)$ with respect to μ_1, Y has density $f_Y(y)$ with respect to μ_2, and their joint density is $f_{(X,Y)}(x, y) = f_X(x)f_Y(y)$. Then X and Y are independent.

It turns out this works in both directions!

F60 If (X, Y) have a joint density that factors as

$$f_{(X,Y)}(x, y) = f_1(x)f_2(y),$$

then the density of X is proportional to $f_1(x)$, and the density of Y is proportional to $f_2(y)$.

CHAPTER 17: JOINT DENSITIES OF RANDOM VARIABLES

E59 Show that X and Y with joint density
$f_{(X,Y)}(x,y) = 2\exp(-x-2y)\mathbb{I}(x \geq 0, y \geq 0)$
are independent.

Answer. Note that the joint density factors into a piece that only depends on the dummy variable for X, and one piece that only depends on the dummy variable for Y. Hence they are independent!

$$f_{(X,Y)}(x,y) = [2\exp(-x)\mathbb{I}(x \geq 0)] \cdot [\exp(-2y)\mathbb{I}(y \geq 0)].$$

ENCOUNTERS

201. JOINT DENSITIES

Suppose (X,Y) have joint density

$$f_{(X,Y)}(x,y) = (12/5)(x^2 + xy)i(x,y)$$

where

$$i(x,y) = \mathbb{I}((x,y) \in [0,1] \times [0,1]).$$

a. Find $\mathbb{P}(X \leq 0.4, Y \leq 0.3)$.
b. Find the density of X, f_X.
c. Are X and Y independent?

202. MORE JOINT DENSITES

Suppose (W,Y) has probability 1 of falling in the set

$$\Omega = [0,2] \times [0,1].$$

Writing $(w,y) \in \Omega$ means that $w \in [0,2]$ and $y \in [0,1]$.

Further, suppose (W,Y) has density

$$f_{(W,Y)}(w,y) = \frac{1}{4}(4 - w - y - wy)\mathbb{I}((w,y) \in \Omega).$$

a. Find $\mathbb{P}(W \leq 1, Y \leq 0.5)$.
b. Find the density of Y.

203. INDEPENDENCE

Let X with density $f_X(s) = \exp(-2s)\mathbb{I}(s \geq 0)$ and Y with density $f_Y(r) = 2r\mathbb{I}(r \in [0,1])$ be independent random variables. What is the joint density

$$f_{(X,Y)}(s,r)?$$

204. MORE INDEPENDENCE

Suppose A, B, and C are independent random variables with densities

$$f_A(s) = 3\exp(-3s)\mathbb{I}(s \geq 0)$$
$$f_B(s) = 4\exp(-4s)\mathbb{I}(s \geq 0)$$
$$f_C(s) = 2.5\exp(-2.5s)\mathbb{I}(s \geq 0).$$

Find $f_{(A,B,C)}(a,b,c)$.

205. MULTIPLE INTEGRALS

Suppose the joint density of X_1 and X_2 is over the region $A = [0,1] \times [0,1]$.

$$f_{(X_1,X_2)}(x_1,x_2) = \frac{3}{8} \cdot \frac{x_1 + x_2}{\sqrt{|x_1 - x_2|}}\mathbb{I}((x_1,x_2) \in A).$$

Find $\mathbb{P}(X_1 > X_2 + 0.1)$ by setting up the integral and then using a numerical solver.

206. RUNNING THE TRIANGLE

Suppose (R,T) has joint density

$$f_{(R,T)}(r,t) = (1/12)(r + 2t)\mathbb{I}((r,t) \in [0,2]^2).$$

What is $\mathbb{P}(T \geq R)$?

207. DISCRETE JOINT DENSITIES

Suppose (X,Y) are uniform over the four points $(-1,1), (-1,-1), (0,0), (1,2)$.
a. What is the density of X?
b. What is the density of Y?
c. Are X and Y independent?

208. Joint Discrete Uniforms

Suppose (X, Y) are uniform over the four points $(-1, 1), (-1, -1), (1, 1), (1, -1)$.
a. What is the density of X?
b. What is the density of Y?
c. Are X and Y independent?

209. Factoring Continuous Densities

Suppose
$f_{(X,Y)}(x, y) = \mathbb{I}(x \in [0, 2]) \exp(-2y)\mathbb{I}(y \geq 0)$.
Show that X and Y are independent.

210. Factoring Discrete Densities

Suppose A and B have joint density $f_{(A,B)}(a, b) = 6a^2 b\mathbb{I}(a \in [0, 1], b \in [0, 1])$.
Prove that A and B are independent.

211. Dependent Joint Continuous Uniforms

Suppose $X = (X_1, X_2)$ is uniform over the upper half of the unit circle given by
$$A = \{(x_1, x_2) : x_2 \geq 0, x_1^2 + x_2^2 \leq 1\}.$$
Since $\text{Leb}(A) = \tau/2$, this has density
$$f_{(X_1, X_2)}(x_1, x_2) = \frac{2}{\tau}\mathbb{I}(\{(x_1, x_2) \in A).$$

Show that X_1 and X_2 are not independent.

212. Joint Continuous Uniforms

Suppose that Ω is in intersection of
$$[0, 1] \times [0, 1] \times [0, 1]$$
and
$$\{(x_1, x_2, x_3) : x_1 + x_2 + x_3 \geq 3/2\}.$$

Suppose $X = (X_1, X_2, X_3)$ has joint density
$$f_X(x) = \frac{6}{5}\mathbb{I}(x \in \Omega).$$

a. What is
$$\mathbb{P}(X_1 < 1/2, X_2 < 1/2, X_3 < 1/2)?$$

b. Find
$$\mathbb{P}(X_1 < 1/2).$$

c. Using the previous parts and considering that
$$\mathbb{P}(X_1 < 1/2) = \mathbb{P}(X_2 < 1/2) = \mathbb{P}(X_3 < 1/2)$$
by symmetry, are X_1, X_2, and X_3 independent?

Chapter 18: Random variables as vector spaces

THE CHEST OF THE MAD KING Triffan had been lost for centuries, and was known to contain thousands of gold and silver pieces. In fact, the Explorer modeled the treasure as being equally likely to contain (0, 0), (0, 1), (0, 2), (1, 2), (1, 1), where (for instance) (1, 2) would mean that there are 1000 gold pieces and 2000 silver pieces. What is the covariance between the number of gold pieces and the number of silver pieces?

Vector space

A *vector space* is built up from two types of objects, *vectors* which can be added together to get new vectors, and *scalars* which can stretch out vectors to give new vectors.

Displacement vectors are typically the first type of vector encountered. In such a vector, the tail of the vector represents the starting location for an object, and the head of the vector shows where the object has been moved to.

While this is a common type of vector, it is far from the only kind. In probability, real-valued random variables can be considered vectors, and then real numbers can be considered scalars.

For instance, if X and Y are random variables, then $X + Y$ is also a random variable. So if these are thought of as vectors, adding them yields another vector.

Similarly, multiplying by a scalar gives something like $4X$ or $-2Y$, which again are random variables.

But that is not the only way to set up vectors! Suppose that X is an integrable random variable, and so has a finite mean. Then the *centered* version of the random variable can be defined.

D69 For an integrable random variable X,
$$X_c = X - \mathbb{E}[X]$$
is the **centered** version of X.

F61 The centered version of an integrable random variable has mean 0.

Proof. The value $\mathbb{E}[X]$ is a constant, and so
$$\mathbb{E}[X - \mathbb{E}[X]] = \mathbb{E}[X] - \mathbb{E}[X] = 0.$$
□

F62 Let X and Y both have mean 0. Then for $a, b \in \mathbb{R}$, $\mathbb{E}[aX + bY] = 0$.

Proof. By linearity of expectations
$$\mathbb{E}[aX + bY] = a\mathbb{E}[X] + b\mathbb{E}[Y] = a \cdot 0 + b \cdot 0 = 0.$$
□

Why is that so important? Because in order to qualify as vectors, linear combinations of vectors must also be vectors.

D70 A **vector space** is a set of vectors V together with a set of scalars S, and two operations with the following properties.
1. There is vector addition, $+$, such that $(\forall v, w \in V)(v + w \in V)$. This addition is associative and commutative. There is a zero vector 0 such that $(\forall v \in V)(v + 0 = v)$. There exist inverses, so that

$(\forall v \in V)(\exists w \in V)(v + w = 0)$.

2. There is scalar multiplication, · such that $(\forall s \in S)(\forall v \in V)(sv \in V)$. This multiplication has an identity element $1 \in S$ so that $(\forall v)(1v = v)$. Also $(\forall a, b \in S)(\forall v \in V)((ab)v = a(bv))$. It is distributive in two ways:

$$(\forall a \in S)(\forall v, w \in V)(a(v + w) = av + aw)$$

and

$$(\forall a, b \in S)(\forall v \in V)((a + b)v = av + bv).$$

It is straightforward to verify that these rules are obeyed by our centered random variables.

F63 Let V be the set of random variables with mean 0, and S be real numbers. Then (V, S) form a vector space.

INNER PRODUCTS

Some vector spaces have what is called an *inner product* that applies to pairs of vectors. For displacement vectors, the value of the inner product tells us about the angle between the two vectors.

For centered random variables, the inner product tells us how the value of one variable affects the average value of the other.

When the inner product is high, a high value of one centered variable leads to a high value for the other. When the inner product is negative, then a high value of one centered variable leads to a lower value for the other.

In general, an *inner product* is defined as follows.

D71 For a vector space (V, S), a **real valued inner product** is a function that maps pairs of vectors to a real number with four properties. For $x, y \in V$, write $\langle x, y \rangle$ for the inner product of x and y. Let x, y, and z be vectors, and α be a scalar. Then the four properties obeyed by an inner product are as follows.
1. $\langle x + y, z \rangle = \langle x, z \rangle + \langle y, z \rangle$.
2. $\langle \alpha x, y \rangle = \alpha \langle x, y \rangle$.
3. $\langle x, y \rangle = \langle y, x \rangle$.
4. $\langle x, x \rangle \geq 0$, where equality holds if and only if $x = 0$.

For displacement vectors, the usual inner product is called the *dot product*. For integrable random variables, the inner product is called the *covariance*.

D72 Given integrable random variables X and Y, if XY is integrable then define the covariance between X and Y as

$$\text{cov}(X, Y) = \mathbb{E}[(X - \mathbb{E}[X])(Y - \mathbb{E}[Y])].$$

NORMS

In mathematics, the term *norm* refers to a measure of the size of a vector. To be a valid size, a norm needs to have three properties. First, if you scale the vector by c, the norm of the vector should grow like c. Second, the triangle inequality holds: the norm of the sum of two vectors is at most the sum of the norms of the vectors. Finally, a norm evaluates to 0 if and only if v is the zero vector.

D73 A **norm** of a vector space takes as input a vector and returns a nonnegative real number. Write the norm of v as v. All norms obey the following rules.
1. For any scalar $c \in \mathbb{R}$ and vector v, $||cv|| = |c| \cdot ||v||$.
2. For any vectors v and w,

$$||v + w|| \leq ||v|| + ||w||.$$

3. For any vector v, $v \geq 0$, and only equals 0 if v is the zero vector.

For vector spaces with an inner product, one easy way to build a norm is to take the square root of the inner product of a vector with itself.

D74 Call $v = \sqrt{\langle v, v \rangle}$ an **inner product norm**.

When the dot product is used with displacement vectors, this gives the *Euclidean norm* which is commonly considered the length of the vector.

If covariance is used as the inner product, the covariance of a random variable with itself is called the *variance*. The square root of this gives the norm called the *standard deviation*.

D75 The **variance** of integrable random variable X is
$$\text{var}(X) = \mathbb{E}[(X - \mathbb{E}[X])(X - \mathbb{E}[X])].$$
The **standard deviation** of random variable X is
$$\text{sd}(X) = \sqrt{\text{var}(X)}.$$

CALCULATING THE COVARIANCE AND STANDARD DEVIATION

The definitions of covariance and standard deviation are not the usual way for actually calculating these values. instead, the following formulas are often used.

F64 For integrable X and Y,
$$\text{cov}(X, Y) = \mathbb{E}[XY] - \mathbb{E}[X]\mathbb{E}[Y]$$
$$\text{var}(X) = \mathbb{E}[X^2] - \mathbb{E}[X]^2$$

These follow from linearity of expectations.

Proof. For covariance:
$$c = \text{cov}(X, Y)$$
$$= \mathbb{E}[(X - \mathbb{E}[X])(Y - \mathbb{E}[Y])]$$
$$= \mathbb{E}[XY - Y\mathbb{E}[X] - X\mathbb{E}[Y] + \mathbb{E}[X]\mathbb{E}[Y]]$$
$$= \mathbb{E}[XY] - \mathbb{E}[X]\mathbb{E}[Y] - \mathbb{E}[X]\mathbb{E}[Y] + \mathbb{E}[X]\mathbb{E}[Y]$$
$$= \mathbb{E}[XY] - \mathbb{E}[X]\mathbb{E}[Y].$$

Applying this to the variance gives
$$\text{var}(X) = \text{cov}(X, X) = \mathbb{E}[XX] - \mathbb{E}[X]\mathbb{E}[X].$$
□

SOLVING THE STORY

The density here is $1/5$ for each of the (X, Y) possible values: $(0, 0)$, $(0, 1)$, $(0, 2)$, $(1, 2)$, or $(1, 1)$.

To find a mean of something like XY, just multiply X times Y times the probabilities that each occurs and add them up! That leads to

$$\mathbb{E}[XY] = 0 \cdot 0 \cdot \frac{1}{5} + 0 \cdot 1 \cdot \frac{1}{5} + \cdots + 1 \cdot 1 \cdot \frac{1}{5}$$
$$\mathbb{E}[X] = 0 \cdot \frac{1}{5} + 0 \cdot \frac{1}{5} + 0 \cdot \frac{1}{5} + 1 \cdot \frac{1}{5} + 1 \cdot \frac{1}{5}$$
$$= \frac{2}{5}.$$
$$\mathbb{E}[Y] = 0 \cdot \frac{1}{5} + 1 \cdot \frac{1}{5} + 2 \cdot \frac{1}{5} + 2 \cdot \frac{1}{5} + 1 \cdot \frac{1}{5} = \frac{6}{5}.$$

This makes the covariance
$$\frac{3}{5} - \frac{2}{5} \cdot \frac{6}{5} = \frac{15 - 12}{25} = \frac{3}{25},$$
which is 0.12.

If X is the number of 1000's of gold pieces, and Y is the number of 1000's of silver, then
$$\text{cov}(X, Y) = 0.12.$$

Therefore, by the rules of covariance,
$$\text{cov}(1000X, 1000Y) = 10^6(0.12) = \boxed{120000}.$$

Proof Covariance is an Inner Product

Here the fact that covariance is an inner product is shown. The properties can be shown one at a time. The first property is that covariance is *distributive*.

F65 Let X, Y, and W be integrable with finite covariance between each pair. Then
$$\operatorname{cov}(X+Y, W) = \operatorname{cov}(X, W) + \operatorname{cov}(Y, W).$$

Proof.
$$\begin{aligned}
C &= \operatorname{cov}(X+Y, W) \\
&= \mathbb{E}[(X+Y)W] - \mathbb{E}[X+Y]\mathbb{E}[W] \\
&= \mathbb{E}[XW + YW] - (\mathbb{E}[X] + \mathbb{E}[Y])\mathbb{E}[W] \\
&= \mathbb{E}[XW] + \mathbb{E}[YW] - \mathbb{E}[X]\mathbb{E}[W] - \mathbb{E}[Y]\mathbb{E}[W] \\
&= \operatorname{cov}(X, W) + \operatorname{cov}(Y, W).
\end{aligned}$$
□

The second property is *scaling*.

F66 Let X and Y have finite covariance, and $a \in \mathbb{R}$. Then
$$\begin{aligned}
\operatorname{cov}(aX, Y) &= \mathbb{E}[(aX)Y] - \mathbb{E}[aX]\mathbb{E}[Y] \\
&= a\mathbb{E}[XY] - a\mathbb{E}[X]\mathbb{E}[Y] \\
&= a\operatorname{cov}(X, Y).
\end{aligned}$$

The third property is commutativity.

F67 For X and Y with finite covariance, $\operatorname{cov}(X, Y) = \operatorname{cov}(Y, X)$.

Proof. Note
$$\begin{aligned}
\operatorname{cov}(X, Y) &= \mathbb{E}[XY] - \mathbb{E}[X]\mathbb{E}[Y] \\
&= \mathbb{E}[YX] - \mathbb{E}[Y]\mathbb{E}[X] \\
&= \operatorname{cov}(Y, X).
\end{aligned}$$
□

The final property says that an inner product is nonnegative, and only equals 0 when the vector is 0. To show this, it is necessary to define our vectors a bit more carefully.

D76 Say that X is **shift equivalent** to Y if there exists a c such that $X + c$ and Y have the same distribution.

Then for us a vector technically consists of a set a shift-equivalent integrable random variables. In particular, all constants are shift equivalent to the number 0.

F68 For all integrable random variables X, it holds that $\operatorname{cov}(X, X) \geq 0$. Moreover, if $\operatorname{cov}(X, X) = 0$, then there exists a constant $c \in \mathbb{R}$ such that $\mathbb{P}(X = c) = 1$.

Proof. The square of any number is nonnegative, hence
$$\operatorname{cov}(X, X) = \mathbb{E}[(X - \mathbb{E}[X])^2] \geq \mathbb{E}[0] = 0.$$

Now assume $\operatorname{cov}(X, X) = 0$. Let $\epsilon > 0$. Then indicator functions are always 0 or 1, so multiplying a nonnegative expression by an indicator function can only possibly make it smaller. Hence
$$\begin{aligned}
0 &= \mathbb{E}[(X - \mathbb{E}[X])^2] \\
&\geq \mathbb{E}[(X - \mathbb{E}[X])^2 \mathbb{I}(|X - \mathbb{E}[X]| \geq \epsilon)].
\end{aligned}$$

For the indicator to be 1, it must hold that $(X - \mathbb{E}[X])^2 \geq \epsilon^2$. Hence
$$\begin{aligned}
0 &\geq \mathbb{E}[(X - \mathbb{E}[X])^2 \mathbb{I}(|X - \mathbb{E}[X]| \geq \epsilon)] \\
&\geq \mathbb{E}[\epsilon^2 \mathbb{I}(|X - \mathbb{E}[X]| > \epsilon)] \\
&= \epsilon^2 \mathbb{P}(|X - \mathbb{E}[X]| > \epsilon) \geq 0
\end{aligned}$$
which means
$$\epsilon^2 \mathbb{P}(|X - \mathbb{E}[X]| > \epsilon) = 0.$$

Since $\epsilon^2 > 0$, that means $\mathbb{P}(|X - \mathbb{E}[X]| > \epsilon) = 0$ for all $\epsilon > 0$.

In particular, letting $\epsilon \in \{1, 1/2, 1/3, \ldots\}$ gives us a countable sequence of events whose total probability is 0. The negation of the union of those events then gives $\mathbb{P}(|X - \mathbb{E}[X]| = 0) = 1$.

That is, $\mathbb{P}(X = \mathbb{E}[X]) = 1$, which completes the proof. □

F69 Covariance is an inner product.

Proof. This follows immediately from the last four facts. □

ENCOUNTERS

213. RULES OF INNER PRODUCTS

Say $\operatorname{cov}(X, Y) = 4.2$. Find $\operatorname{cov}(3X, -2Y)$.

214. MORE RULES OF INNER PRODUCTS

Suppose $\operatorname{cov}(X, Y) = 4.2$ and $\operatorname{cov}(X, W) = -2.3$. What is $\operatorname{cov}(2X, Y - W)$?

215. RULES OF INNER PRODUCT NORMS

Suppose that $\operatorname{sd}(X) = 1.8$.
a. What is the variance of X?
b. What is $\operatorname{sd}(3X)$?
c. What is $\operatorname{sd}(-3X)$?

216. STANDARD DEVIATION

Suppose $\operatorname{sd}(Y) = 0.4$.
a. What is $\operatorname{var}(Y)$?
b. What is $\operatorname{sd}(Y - Y)$?
c. What is $\operatorname{sd}(Y + Y)$?

217. STANDARD DEVIATION OF DISCRETE RANDOM VARIABLES

Say X is discrete with density $f_X(-1) = 0.6$, $f_X(0) = 0.3$, $f_X(1) = 0.1$.

a. Find $\mathbb{E}[X]$.
b. Find $\operatorname{sd}(X)$.

218. DISCRETE VIA DENSITY

Say W is discrete with density $f_W(0) = f_W(1) = 0.2$, $f_W(2) = f_W(3) = 0.3$.
a. Find $\mathbb{E}[W]$.
b. Find $\operatorname{sd}(W)$.

219. COVARIANCE OF A JOINT UNIFORM

Let (X, Y) be uniform over the triangle with vertices $(0, 0)$, $(0, 1)$, and $(1, 0)$. Find $\operatorname{cov}(X, Y)$.

220. A UNIFORM TRIANGLE

Suppose (A, B) is uniform over the triangle with vertices $(0, 0), (1, 0), (1, 1)$. Find $\operatorname{cov}(A, B)$.

221. CONTINUOUS UNIFORM PART I

Suppose (X, Y) is uniform over the triangle A given by

$$A = \{(x, y) : 0 \le x \le 1, 0 \le y \le 1, x \le y\},$$

so has density

$$f_{(X,Y)}(x, y) = 2\mathbb{I}((x, y) \in A).$$

a. Find $\mathbb{E}[XY]$.
b. Find $\mathbb{E}[X]$.
c. Find $\mathbb{E}[Y]$.
d. Find $\operatorname{cov}(X, Y)$.

222. CONTINUOUS UNIFORM PART II

Suppose (S, T) is uniform over the triangle B given by

$$B = \{(s, t) : 0 \le x \le 1, 0 \le y \le 2, 2x \le y\},$$

so has density

$$g_{(S,T)}(s, t) = 2\mathbb{I}((s, t) \in B).$$

a. Find $\mathbb{E}[ST]$.
b. Find $\mathbb{E}[S]$.
c. Find $\mathbb{E}[T]$.
d. Find $\mathrm{cov}(S, T)$.

Chapter 19: Correlation

THE CHEST OF THE MAD KING Triffan had been lost for centuries, and was known to contain thousands of gold and silver pieces. In fact, the Explorer modeled the treasure as being equally likely to contain (0, 0), (0, 1), (0, 2), (1, 2), (1, 1), where (for instance) (1, 2) would mean that there are 1000 gold pieces and 2000 silver pieces. What is the correlation between the number of gold pieces and the number of silver pieces?

has units equal to the square of the units of X. Hence $\mathrm{sd}(X) = \sqrt{\mathrm{var}(X)}$ has units equal to the units of X. Finally, that means that $X/\mathrm{sd}(X)$ is unitless, it does not have any units at all!

If both X and Y are centered and then inversely scaled by their standard deviation, the result is unitless. The covariance of these scaled random variables is called the *correlation* of the original random variables. Remember that scaling can be pulled out of an inner product, which gives rise to the following.

> **D77** The **correlation** between integrable random variables X and Y with finite covariance and nonzero standard deviations is
> $$\mathrm{cor}(X, Y) = \frac{\mathrm{cov}(X, Y)}{\mathrm{sd}(X)\,\mathrm{sd}(Y)}.$$

Correlation

In the past chapter the idea of *covariance* was introduced. Covariance is a type of product, an inner product to be exact, between two random variables X and Y. So because it is like the product of two things, the units of covariance will be the product of the units of X and Y. For instance, if X is measures in miles and Y is measured in per hour units, then $\mathrm{cov}(X, Y)$ has units of miles per hour.

It is often useful to have a measurement of the relation between X and Y that is *unitless*. For instance, the angle formed by two displacement vectors v and w is such a measurement: it is unaffected by the length of the individual vectors.

To accomplish this with random variables, think about the following. First, since $\mathrm{var}(X) = \mathrm{cov}(X, X)$, $\mathrm{var}(X)$

Solving the Story

To calculate the correlation between two random variables X and Y, it is necessary to calculate $\mathbb{E}[X]$, $\mathbb{E}[Y]$, $\mathbb{E}[X^2]$, $\mathbb{E}[Y^2]$, and $\mathbb{E}[XY]$. For the question of the day, the probability of each outcome is $1/5$. Then sum the values of any function $g(X, Y)$ over the five possibilities $(0,0), (0,1), (0,2), (1,2), (1,1)$.

$$\mathbb{E}[X] = \frac{1}{5}[0 + 0 + 0 + 1 + 1] = \frac{2}{5}$$

$$\mathbb{E}[Y] = \frac{1}{5}[0 + 1 + 2 + 2 + 1] = \frac{6}{5}$$

$$\mathbb{E}[X^2] = \frac{1}{5}[0^2 + 0^2 + 0^2 + 1^2 + 1^2] = \frac{2}{5}$$

$$\mathbb{E}[Y^2] = \frac{1}{5}[0 + 1^2 + 2^2 + 2^2 + 1^2] = \frac{10}{5}$$

$$\mathbb{E}[XY] = \frac{1}{5}[0 \cdot 0 + 0 \cdot 1 + 0 \cdot 1 + 0 \cdot 2 + 1 \cdot 2 + 1 \cdot 1]$$
$$= \frac{3}{5}$$

That makes

$$\mathrm{cor}(X,Y) = \frac{(3/5) - (2/5)(6/5)}{\sqrt{[(2/5) - (2/5)^2][(10/5) - (6/5)^2]}},$$

which is about $\boxed{0.3273}$.

BOUNDS ON THE CORRELATION

Correlation measures how aligned the random variables are. For instance, suppose $Y = 2X$. Then

$$\begin{aligned}\mathrm{cor}(X, 2X) &= \frac{\mathrm{cov}(X, 2X)}{\mathrm{sd}(X)\,\mathrm{sd}(2X)} \\ &= \frac{2\,\mathrm{var}(X)}{2\,\mathrm{sd}(X)\,\mathrm{sd}(X)} \\ &= 1.\end{aligned}$$

Note the 2 canceled there. That will happen for any positive constant. Now try X times a negative constant.

$$\begin{aligned}\mathrm{cor}(X, -2X) &= \frac{\mathrm{cov}(X, -2X)}{\mathrm{sd}(X)\,\mathrm{sd}(-2X)} \\ &= \frac{-2\,\mathrm{var}(X)}{|-2|\,\mathrm{sd}(X)\,\mathrm{sd}(X)} \\ &= -1.\end{aligned}$$

So the answer was 1 for the correlation of a random variable and the random variable times a positive constant, and -1 between the random variable and the random variable times a negative constant. It turns out that this is the largest and smallest that the correlation can be. This is due to a general fact about inner products.

Recall that the *inner product norm* of v for a given inner product is $\sqrt{\langle v, v \rangle}$. The Cauchy Schwartz inequality says that the absolute value of an inner product between v and w can be at most the product of the norm of v times the norm of w. Moreover, equality is only reached when one vector is a multiple of the other.

T7 The Cauchy Schwartz inequality
For a vector space with an inner product, for any two vectors v and w,

$$|\langle v, w \rangle| \le \sqrt{\langle v, v \rangle \cdot \langle w, w \rangle}.$$

In addition, equality holds if and only if there exist real numbers α and β, at least one of which is nonzero, such that $\alpha v + \beta w = 0$.

The proof is at the end of this chapter.

UNCORRELATED

For displacement vectors, having a dot product of zero means that the vectors are *perpendicular* or *orthogonal* to one another.

When $\mathrm{cor}(X, Y) = 0$, call the random variables *uncorrelated*.

D78 Two random variables X and Y are *uncorrelated* if their correlation exists and is 0.

Correlation is about how the average value of variables relate to each other. Independence is stronger than uncorrelated, as independence says that knowing one variable has no effect on the distribution of the other variable. That is, X and Y independent means $[X \mid Y] \sim [X]$.

F70 Suppose that X and Y are independent and have a correlation. Then the correlation is 0.

Proof. Suppose that X and Y are independent. Then

$$\begin{aligned}\mathbb{E}[XY] &= \mathbb{E}[\mathbb{E}[XY \mid Y]] \\ &= \mathbb{E}[Y\mathbb{E}[X \mid Y]] \\ &= \mathbb{E}[Y\mathbb{E}[X]] \\ &= \mathbb{E}[X]\mathbb{E}[Y].\end{aligned}$$

So their covariance is 0. □

In other words, independent random variables are uncorrelated. However, the converse is not always true: random variables can be uncorrelated but still dependent.

E60 Suppose (X, Y) is uniformly drawn from $(-1, -2)$, $(-1, 2)$, $(1, -1)$, $(1, 1)$. Show that (X, Y) are not independent, but are uncorrelated.
Answer. Intuitively, they are dependent because knowing if X is -1 or 1 changes the distribution of Y. They are uncorrelated because knowing if X is -1 or 1 does not change the fact that the average value of Y is 0.

Formally,
$$\mathbb{P}(X = -1, Y = -1) = 0,$$
but
$$\mathbb{P}(X = -1)\mathbb{P}(Y = -1) = (1/2)(1/4) = 1/8$$
so X and Y are dependent.

For correlation:
$$\mathbb{E}[X] = \frac{1}{4}[-1 + -1 + 1 + 1] = 0,$$
$$\mathbb{E}[Y] = \frac{1}{4}[-2 + 2 + -1 + 1] = 0,$$
$$\mathbb{E}[XY] = \frac{1}{4}[(-1)(-2) + (-1)(2) + (1)(-1) + (1)(1)] = 0,$$
so $\mathbb{E}[XY] - \mathbb{E}[X]\mathbb{E}[Y] = 0 - (0)(0) = 0$. Hence they are uncorrelated.

PROPERTIES OF VARIANCE AND COVARIANCE.

Many inner product properties carry over to variance and covariance. The first is the extension of the Pythagorean theorem to non-orthogonal vectors.

F71 For any inner product norm
$$||v_1 + \cdots + v_n||^2 = ||v_1||^2 + \cdots + ||v_n||^2 + \sum_{i \neq j} \langle v_i, v_j \rangle.$$

Proof. This follows from the distributive rules
$$||v_1 + \cdots + v_n||^2 = \langle v_1 + \cdots + v_n, v_1 + \cdots + v_n \rangle$$
$$= \sum_{i=1}^{n} \langle v_i, v_i \rangle + \sum_{i \neq j} \langle v_i, v_j \rangle$$
$$= \sum_{i=1}^{n} v_i^2 + \sum_{i \neq j} \langle v_i, v_j \rangle.$$
□

In terms of variance and covariance, this means the following.

F72 For $X_1, \ldots, X_n$ random variables with finite variance where each pair has finite covariance,
$$\text{var}(X_1 + \cdots + X_n) = \sum_{i=1}^{n} \text{var}(X_i) + 2\sum_{i<j} \text{cov}(X_i, X_j).$$

When the inner products $\langle v_i, v_j \rangle$ are 0, this gives the *Pythagorean theorem.*

T8 Pythagorean Theorem
If $\langle v_i, v_j \rangle = 0$ for all $i \neq j$ in $\{1, \ldots, n\}$, then
$$||v_1 + \cdots + v_n||^2 = ||v_1||^2 + \cdots + ||v_n||^2.$$

In Euclidean space this is often called the *distance formula*. In probability $X^2 = \text{var}(X)$, so it gives the following.

F73 If each pair in $X_1, \ldots, X_n$ is uncorrelated, then
$$\text{var}(X_1 + \cdots + X_n) = \text{var}(X_1) + \cdots + \text{var}(X_n),$$
or in terms of standard deviation
$$\text{sd}(X_1 + \cdots + X_n)^2 = \text{sd}(X_1)^2 + \cdots + \text{sd}(X_n)^2.$$

Remember that independent random variables are always uncorrelated, so the theorem applies to the sum of independent random variables. In addition, scaling a vector changes the standard deviation by a factor equal to the absolute value of the scale, which gives the following fact about sample averages.

> **F74** If $X_1, \ldots, X_n$ are iid distributed as X with finite standard deviation, then
> $$\operatorname{sd}\left(\frac{X_1 + \cdots + X_n}{n}\right) = \frac{\operatorname{sd}(X)}{\operatorname{sqrt}(n)}.$$

Proof. Using scaling and the Pythagorean Theorem:

$$\begin{aligned}
\operatorname{sd}\left(\frac{X_1 + \cdots + X_n}{n}\right) &= \frac{1}{n}\operatorname{sd}(X_1 + \cdots + X_n) \\
&= \frac{1}{n}\sqrt{\operatorname{sd}(X_1 + \cdots X_n)^2} \\
&= \frac{1}{n}\sqrt{\operatorname{sd}(X_1)^2 + \cdots + \operatorname{sd}(X_n)^2} \\
&= \frac{1}{n}\sqrt{n\operatorname{sd}(X_1)^2} \\
&= \frac{\sqrt{n}\operatorname{sd}(X)}{n} \\
&= \frac{\operatorname{sd}(X)}{\sqrt{n}}.
\end{aligned}$$

□

PROOF OF THE CAUCHY SCHWARTZ INEQUALITY.

The Cauchy Schwartz inequality says that the absolute value of the inner product is at most the product of the inner product norm of two vectors. Equality holds only when the two vectors are linearly dependent, that is, when there exist scalars α and β (not both zero) such that $\alpha v + \beta w = 0$.

In probability this says that
$$|\operatorname{cov}(X, Y)| \leq \operatorname{sd}(X)\operatorname{sd}(Y),$$
with equality when $X = \alpha Y$ or $Y = \alpha X$ for some α.

The proof given holds for general inner products.

Proof. If $w = 0$ then both sides are zero. Equality holds and v and w are linearly dependent.

Assume $w \neq 0$. Then $\langle w, w \rangle > 0$, so let
$$\lambda = \frac{\langle v, w \rangle}{\langle w, w \rangle}.$$

By the positive definite property of the inner product:
$$0 \leq \langle v - \lambda w, v - \lambda w \rangle$$

Using linearity and symmetry allows us to multiply out the right hand side to get

$$\begin{aligned}
0 &\leq \langle v, v \rangle - 2\lambda \langle v, w \rangle + \lambda^2 \langle w, w \rangle \\
&= \langle v, v \rangle - 2\frac{\langle v, w \rangle^2}{\langle w, w \rangle} + \frac{\langle v, w \rangle^2 \langle w, w \rangle}{\langle w, w \rangle^2} \\
&= \langle v, v \rangle - \frac{\langle v, w \rangle^2}{\langle w, w \rangle}
\end{aligned}$$

$$\langle v, w \rangle^2 \leq \langle v, v \rangle \langle w, w \rangle$$
$$|\langle v, w \rangle| \leq \sqrt{\langle v, v \rangle \langle w, w \rangle}.$$

If equality holds, then either $w = 0$ (in which case $(0)(v) + (1)(w) = 0$) or $\langle v - \lambda w, v - \lambda w \rangle = 0$. In the latter case $v - \lambda w = 0$ so v and w are also linearly dependent. □

ENCOUNTERS

223. AFFINE TRANSFORMS

Suppose X and Y have correlation 0.4276. What is the correlation between $2X + 4$ and $5Y + 3$?

224. A BIT OF NEGATIVITY

Suppose W_1 and W_2 have correlation 0.3. What is the correlation between W_1 and $-W_2$?

225. Correlation of Discrete Uniforms

For $(X,Y) \sim \mathsf{Unif}(\{(0,0),(2,0),(2,1)\})$, find the correlation between X and Y.

226. Correlation of Independent Random Variables

Suppose (X,Y) has density $f_{X,Y}(x,y) = 2\exp(-2x-y)\mathbb{I}(x,y \geq 0)$. Find $\mathrm{cor}(X,Y)$.

227. Understanding Functions

Suppose $U \sim \mathsf{Unif}([0,1])$. Find $\mathrm{cor}(U,U^2)$.

228. Another Negative Result

For $U \sim \mathsf{Unif}([0,1])$, find $\mathrm{cor}(U,1-U)$.

229. The Pythagorean Theorem

Suppose S,T,R are independent random variables with variances of $1.1, 2.8, 0.6$ respectively.
a. What is $\mathrm{var}(S+T+R)$?
b. What is $\mathrm{var}(S-2T)$?

230. More Pythagoras

Suppose that $X_1,\ldots,X_n$ are independent, and $\mathrm{var}(X_i) = 1/i$. Find $\mathrm{var}(X_1 + \cdots + X_6)$.

231. Sample Averages

Suppose X has standard deviation 3.2. What is the standard deviation of $(X_1 + \cdots + X_{100})/100$ if the X_i are iid X?

232. How Many to Average?

Suppose Y has standard deviation 0.4. How big does n need to be for the standard deviation of $(Y_1 + \cdots + Y_n)/n$ to be at most 0.01?

Chapter 20: Adding random variables

The King needed for the Defender of the Realm to return to the capital. Quickly a Messenger was sent to retrieve the Defender. If it took time $T \sim \mathsf{Exp}(1.2)$ days for the Messenger to find the Defender, and then time $R \sim \mathsf{Exp}(1.5)$ for the Defender to return to the capital, what is the chance that it would take more than 2 days for the Defender to reach the capital?

Adding discrete random variables

Suppose that $X \sim \mathsf{d4}$ and $Y \sim \mathsf{d6}$ are independent rolls of two dice. What is the chance that $X + Y = 7$? Well, there are four outcomes of X and Y that sum to 7:

$$(X, Y) \in \{(1,6), (2,5), (3,4), (4,3)\}.$$

Each of the outcomes are independent, and there are 24 total possibilities for (X, Y), so there is a 4 out of 24 or $1/6$ chance that $X + Y = 7$.

Another way to think about this problem is through the densities of the random variables. No matter what Y is chosen, the only way to add to 7 is for $X = 7 - Y$. Hence

$$\mathbb{P}(X + Y = 7) = \sum_y \mathbb{P}(X + Y = 7, Y = y)$$
$$= \sum_y \mathbb{P}(X = 7 - y, Y = y)$$
$$= \sum_y f_{(X,Y)}(7 - y, y)$$
$$= \sum_y f_X(7 - y) f_Y(y).$$

This is a sum, which is the same as an integral in counting measure:

$$\mathbb{P}(X + Y = 7) = \int_y f_Y(y) f_X(7 - y) \, d\#(y).$$

Of course, instead of summing over $Y = y$, the problem could have been solved by summing over $X = x$, yielding,

$$\mathbb{P}(X + Y = 7) = \int_x f_X(x) f_Y(7 - x) \, d\#(x).$$

It turns out that this way of calculating probabilities involving sums of random variables works for all densities, for all measures, not just counting measures. This type of sum is called a *convolution*.

> **T9** For random variables X and Y with joint density $f_{(X,Y)}(x, y)$ with respect to $\mu_X \times \mu_Y$, the density of $S = X + Y$ is
> $$f_S(s) = \int_x f_{(X,Y)}(x, s - x) \, d\mu_X(x)$$
> $$= \int_y f_{(X,Y)}(s - y, y) \, d\mu_Y(y).$$

Proof. Let X and Y be random variables with joint density $f_{(X,Y)}$ with respect to $\mu_X \times \mu_Y$. Then by the Fundamental Theorem of Probability, for any measurable set A,

$$\begin{aligned}\mathbb{P}(X+Y \in A) &= \mathbb{E}[\mathbb{I}(X+Y \in A)] \\ &= \mathbb{E}[\mathbb{E}[\mathbb{I}(X+Y \in A)|X]] \\ &= \mathbb{E}[\mathbb{E}[\mathbb{I}(Y \in A-X)|X]],\end{aligned}$$

where for any c, $A - c = \{r : r + c \in A\}$. Continuing, this gives,

$$\begin{aligned}\mathbb{P}(X+Y \in A) &= \mathbb{E}_X\left[\int_{y \in A-X} f_{(X,Y)}(X,y) d\mu_Y\right] \\ &= \int_x \int_{y \in A-x} f_{(X,Y)}(x,y)\, d\mu_Y\, d\mu_X \\ &= \int_x \int_{s=x+y, s \in A} f_{(X,Y)}(x,y)\, d\mu_Y\, d\mu_X\end{aligned}$$

In the interior integral $y = s - x$ is just a shifting of y, $d\mu_Y = d\mu_S$, giving

$$\mathbb{P}(X+Y \in A) = \int_x \int_{s \in A} f_{(X,Y)}(x, s-x)\, d\mu_S\, d\mu_X$$

Because the joint density is nonnegative, Tonelli's theorem allows us to swap the order of the integrated integrals to give

$$\mathbb{P}(X+Y \in A) = \int_{s \in A} \int_x f_{(X,Y)}(x, s-y)\, d\mu_X\, d\mu_S$$

But that means that the inside integral is the density of $S = X + Y$.

The other equation just conditions on Y inside the expected value rather than X near the beginning. □

INDEPENDENT RANDOM VARIABLES

When the random variables X and Y are independent, then the joint density is the product of the marginal densities so

$$f_{(X,Y)}(x,y) = f_X(x)f_Y(y).$$

In general, for real valued functions f and g with respect to measure μ, the *convolution* of the functions is defined as follows.

D79 The **convolution** of real valued functions f and g with respect to measure μ is

$$[f * g](s) = \int_x f(x)g(s-x)\, d\mu$$

Then the density of the sum of independent random variables is just the convolution of the random variables' densities.

F75 If X and Y are independent random variables with densities f_X and f_Y with respect to measure μ, then the density of $S = X + Y$ is $f_X * f_Y$.

SOLVING THE STORY

In the story, the first time is $T \sim \text{Exp}(1.2)$, with density $f_T(t) = 1.2\exp(-1.2t)\mathbb{I}(t \geq 0)$. Similarly, $R \sim \text{Exp}(1.5)$, so $f_R(r) = 1.5\exp(-1.5r)\mathbb{I}(r \geq 0)$. Therefore, the convolution

$$[f_T * f_R](s)$$

is the integral

$$\int_t g_1(t)\mathbb{I}(t \geq 0)g_2(s-t)\mathbb{I}(s-t \geq 0)\, dt$$

where

$$\begin{aligned}g_1(t) &= 1.2\exp(-1.2t) \\ g_2(s-t) &= 1.5\exp(-1.5(s-t))\end{aligned}$$

This integral was written to emphasize the role of indicator functions in the individual densities. Recall that the product of indicator functions is the logical and of the arguments. So

$$\mathbb{I}(t \geq 0)\mathbb{I}(s-t \geq 0) = \mathbb{I}(0 \leq t \leq s).$$

That makes the convolution

$$\int_t 1.8\exp(-1.5s + 0.3t)\mathbb{I}(0 \leq t \leq s)\, dt.$$

When $s < 0$ then $\mathbb{I}(0 \leq t \leq s) = 0$ for all t and the integral is 0.

When $s \geq 0$, then the indicator can be used to set the limits of the integral, giving

$$[f_T * f_R](s) = \int_{t=0}^{s} 1.8 \exp(-1.5s + 0.3t) \, dt$$
$$= 6 \exp(-1.5s + 0.3t)|_0^s$$
$$= 6[\exp(-1.2s) - \exp(-1.5s)].$$

Therefore, the density of $T + R$ at s is

$$6[\exp(-1.2s) - \exp(-1.5s)]\mathbb{I}(s \geq 0).$$

This density looks like this:

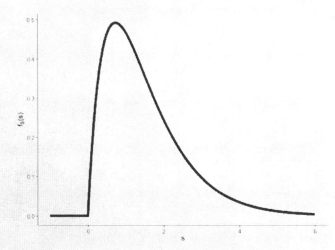

This looks roughly like the shape of a gamma distribution, but it is actually different.

To finish the problem, this density is integrated from 2 to infinity.

$$\mathbb{P}(T + R \geq 2)$$
$$= \int_{s \geq 2} 6[\exp(-1.2s) - \exp(-1.5s)]\mathbb{I}(s \geq 0) \, ds$$
$$= 6\left[\frac{\exp(-1.2s)}{-1.2} - \frac{\exp(-1.5s)}{-1.5}\right]\Big|_2^{\infty}$$
$$\approx \boxed{0.2544}.$$

Let's do a discrete example where two discrete uniform random variables are added together. In this case, the trickiest part of the convolution is keeping track of the indicator functions in the densities.

E61 Suppose $X \sim d4$ and $Y \sim d10$ are independent. What is the density of $X + Y$?
Answer. This is

$$f_{X+Y}(s)$$
$$= \int_i f_X(i) f_Y(s - i) \, d\#$$
$$= \sum_i f_X(i) f_Y(s - i)$$
$$= \sum_i \frac{\mathbb{I}(i \in \{1,2,3,4\})}{4} \cdot \frac{\mathbb{I}(s - i \in \{1,2,\ldots,10\})}{10}$$

Note that if $s - i = 7$, then $i = s - 7$. So

$$f_{X+Y}(s)$$
$$= \sum_i \frac{\mathbb{I}(i \in \{1,2,3,4\}, i \in \{s-1, \ldots, s-10\})}{40}$$
$$= \sum_i \frac{\mathbb{I}(i \in \{1,2,3,4\} \cap \{s-10, \ldots, s-1\})}{40}$$

Since each entry in the sum is either $1/40$ or $0/40$, the sum will be the number of integers i that lie in $\{1,2,3,4\} \cap \{s - 10, s - 9, \ldots, s - 1\}$. Or equivalently, $i \in \mathbb{Z}$ such that $\max(1, s - 10) \leq i \leq \min(4, s - 1)$.

Break it down by cases.

case	$\max(1, s-10)$	$\min(4, s-1)$
$s < 2$	1	$s - 1 < 1$
$s < 5$	1	$s - 1$
$5 \leq s \leq 11$	1	4
$11 < s \leq 14$	$s - 10$	4
$s > 14$	$s - 10 > 4$	4

From the table, the density is:

$$\boxed{f_{X+Y}(s) = (s - 1) \cdot \mathbb{I}(s \in \{2, 3, 4\}) + 4 \cdot \mathbb{I}(s \in \{5, \ldots, 11\}) + (14 - s) \cdot \mathbb{I}(s \in \{12, 13, 14\}).}$$

Using Polynomials to Add Discrete Random Variables

A faster way to calculate densities like in the previous example is to use polynomial multiplication. For instance, consider the *generating function* associated with $A \sim d3$:

$$\text{gf}_A(x) = (1/3)x + (1/3)x^2 + (1/3)x^3.$$

There is an x^i term for each i such that $\mathbb{P}(A = i) > 0$, and the coefficient of the term is $\mathbb{P}(A = i)$. Because $A \in \{1, 2, 3\}$, this generating function is a degree 3 polynomial.

For $B \sim d4$, the polynomial is

$$\text{gf}_B(x) = (1/4)x + (1/4)x^2 + (1/4)x^3 + (1/4)x^4.$$

Now let's multiply these two polynomials together. After simplifying, $\text{gf}_A(x)\,\text{gf}_B(x)$ will be

$$\frac{1}{12}x^2 + \frac{2}{12}x^3 + \frac{3}{12}x^4 + \frac{3}{12}x^5 + \frac{2}{12}x^6 + \frac{1}{12}x^7$$

This polynomial is the one associated with $S = A + B$! To see why, consider the x^3 term. This term comes from multiplying the following terms of gf_A and gf_B together:

$$\frac{1}{3}x^1 \cdot \frac{1}{4}x^2 + \frac{1}{3}x^2 \cdot \frac{1}{4}x^1.$$

Because the way that the x^i terms are created,

$$c_s x^s = \sum_{i,j : i+j = s} a_i x^i b_j x^j,$$

which exactly gives the polynomial for the sum.

Now note that this polynomial can be viewed as an expected value:

$$\text{gf}_X(x) = \sum_i \mathbb{P}(X = i) x^i = \mathbb{E}[x^X].$$

D80 The **generating function** of a random variable X is

$$\text{gf}_X(x) = \mathbb{E}[x^X].$$

F76 If X and Y are independent, then

$$\text{gf}_{X+Y}(x) = \text{gf}_X(x) \cdot \text{gf}_Y(x).$$

Proof. Since X and Y are independent, then so are x^X and x^Y. Hence

$$\mathbb{E}[x^{X+Y}] = \mathbb{E}[x^X x^Y] = \mathbb{E}[x^X]\mathbb{E}[x^Y],$$

which completes the proof. □

You can find the generating function for continuous random variables as well.

E62 Suppose $T \sim \text{Exp}(\lambda)$. What is $\text{gf}_T(x)$?
Answer. Use the formula for expectation of a function of a random variable, and assuming $x > 0$,

$$\text{gf}_T(x) = \mathbb{E}[x^T]$$
$$= \int_{s \in \mathbb{R}} x^s \lambda \exp(-\lambda s) \mathbb{I}(s \geq 0)\,ds$$
$$= \int_{s \geq 0} \exp(s \ln(x)) \exp(-\lambda s)\,ds$$
$$= \int_{s \geq 0} \lambda \exp(-(\lambda - \ln(x))s)\,ds$$
$$= -\frac{\lambda}{\lambda - \ln(x)} \exp(-(\lambda - \ln(x))s)\Big|_0^\infty$$
$$= \frac{\lambda}{\lambda - \ln(x)},$$

as long as $\ln(x) < \lambda$, so $x < e^\lambda$.

An important fact is that if two random variables have the same moment generating function for any interval of nonzero length, then they have the same distribution.

F77 Suppose $a < b$ are real numbers, and for all $x \in [a, b]$, $\text{gf}_X(x) = \text{gf}_Y(x)$. Then $X \sim Y$.

E63 Prove using generating functions that for A_1, A_2 are iid $\mathsf{Exp}(\lambda)$, the density of $A_1 + A_2$ is
$$\frac{1}{2}\lambda^2 w \exp(-\lambda w)\mathbb{I}(w \geq 0).$$

Answer. Because A_1 and A_2 are independent,
$$2\,\mathrm{gf}_{A_1+A_2}(x) = \mathrm{gf}_{A_1}(x)\,\mathrm{gf}_{A_2}(x) = \left(\frac{\lambda}{\lambda - \ln(x)}\right)^2,$$
for $x \in [0, \exp(\lambda))$.

Suppose W has density $f_W(w) = \lambda^2 w \exp(-\lambda w)/2\,\mathbb{I}(w \geq 0)$. Then
$$\mathrm{gf}_W(x) = \mathbb{E}[x^W]$$
$$= \int_{w \in \mathbb{R}} x^w \lambda^2 w \exp(-\lambda w)/2\,\mathbb{I}(w \geq 0)\, dw$$
$$= \int_{w \geq 0} \exp(w \ln(x))\lambda^2 w \exp(-\lambda w)/2\, dw$$
$$= \lambda^2 \int_{w \geq 0} w \left[\frac{\exp(-(\lambda - \ln(x))w)}{-(\lambda - \ln(x))}\right]' dw$$

Recall that integration by parts allows us to slide the derivative from one part over to the other part using
$$\int_a^b f(w)g'(w)\, dw = f(w)g(w)\big|_a^b - \int_a^b f'(w)g(w)\, dw.$$

In our case, $f(w) = w$ so $f'(w) = 1$. Also our $g(w)$ is an exponential with a negative leading exponent for $\lambda \geq \ln(x)$. Hence $\lim_{w \to \infty} g(w) = 0$. Also, $f(0) = 0$, so
$$\mathrm{gf}_W(x) = -\lambda^2 \int_{w \geq 0} \frac{\exp(-(\lambda - \ln(x))w)}{-(\lambda - \ln(x))}\, dw$$
$$= \lambda^2 \frac{\exp(-(\lambda - \ln(x))w)}{-(\lambda - \ln(x))^2}\bigg|_0^\infty$$
$$= \frac{\lambda^2}{(\lambda - \ln(x))^2},$$

which is exactly the generating function of $A_1 + A_2$. Hence $W \sim A_1 + A_2$, and the density is correct.

ENCOUNTERS

233. ADDING RANDOM VARIABLES

Suppose $(X, Y) \sim \mathsf{Unif}(\{(1,2), (1,3), (2,2)\})$. What is the density of $X + Y$?

234. MORE ADDITION

Suppose (X, Y) are uniform over $\{(0,0), (0,1), (0,2), (1,0), (1,1), (2,0)\}$. What is the density of $X + Y$?

235. ADDING INDEPENDENT DICE

Suppose $A \sim \mathsf{d4}$ and $B \sim \mathsf{d4}$ are independent. Use the convolution of the densities of A and B to find the density of $A + B$.

236. MORE DICE

If $X \sim \mathsf{d3}$ and $Y \sim \mathsf{d6}$ are independent, find the density of $X + Y$ using the convolution method.

237. ADDING CONTINUOUS RANDOM VARIABLES

Suppose $A \sim \mathsf{Unif}([0,1])$ and $B \sim \mathsf{Exp}(1)$. What is the density of $A + B$?

238. MORE CONTINUOUS ADDITION

Suppose $A \sim \mathsf{Unif}([0,2])$ and $B \sim \mathsf{Exp}(1.2)$. What is the density of $A + B$?

239. GENERATING FUNCTION OF A POISSON

Suppose $X \sim \mathsf{Pois}(\mu)$. Prove that the generating function of X is $\exp(-\mu(1-x))$.

240. Adding Poissons

Suppose $X_1 \sim \text{Pois}(\mu_1)$ and $X_2 \sim \text{Pois}(\mu_2)$ are independent. Prove that $X_1 + X_2 \sim \text{Pois}(\mu_1 + \mu_2)$.

241. Wolfram Alpha to the Rescue

Suppose $X_1, \ldots, X_{10}$ are iid d4. Then using Wolfram Alpha to perform the polynomial multiplications, find the probability that $X_1 + \cdots + X_{10} = 23$.

242. Adding lots of dice

Suppose $X_1, \ldots, X_{10}$ are iid d6. Then using Wolfram Alpha to perform the polynomial multiplications, find the probability that the sum of the X_i are at least 55.

Chapter 21: The Central Limit Theorem

Taxes for the year came from five different regions of the kingdom. If the taxes from each region are independent and modeled as $\mathsf{Unif}([0, 4000])$, is it possible to estimate the chance that the taxes collected are at least 14,000?

The Moment Generating Function

The generating function allows us to prove things about the sum of random variables much more quickly than the convolution method. By modifying this function slightly, there is another unexpected benefit.

> **D81** The **moment generating function** of a random variable X is
> $$\mathrm{mgf}_X(t) = \mathbb{E}[\exp(tX)].$$

The moment generating function has the same information as the generating function for $s > 0$. This is because
$$\mathrm{gf}_X(s) = \mathrm{mgf}_X(\ln(s)).$$

By definition, $\mathrm{mgf}_X(0) = 1$, but for $t \neq 0$, there is no guarantee that this expected value is finite.

Why do this? One reason is that another word for the expected value of a random variable is the *first moment*.

> **D82** For an integrable random variable X, the **first moment** (or often just **moment**) of X is $\mathbb{E}[X]$.
>
> Similarly, if X^i is integrable, then the ith moment of X is just $\mathbb{E}[X^i]$.

Now consider again $\mathrm{mgf}_X(t) = \mathbb{E}[\exp(tX)]$. Suppose the derivative of the mgf with respect to t exists. If the derivative could be swapped with the expectation operator, that would give the following:

$$\frac{d}{dt}\mathbb{E}[\exp(tX)] = \mathbb{E}\left[\frac{d\exp(tX)}{dt}\right] = \mathbb{E}[X\exp(tX)].$$

Now, swapping derivatives and means is not always valid, but it turns out to be valid for a moment generating function that is finite for a positive length interval that contains $t = 0$.

Now plug in 0 for t in $\mathbb{E}[X\exp(tX)]$. That leaves us with $\mathbb{E}[X]$. In other words, the first moment of X is the first derivative of $\mathrm{mgf}_X(t)$ evaluated at $t = 0$.

By taking more derivatives of t, it is possible to obtain higher moments of X.

> **F78** Suppose $\mathrm{mgf}_X(t)$ exists for a positive length interval. Then for a positive integer i,
> $$\left.\frac{d^i}{dt^i}\mathrm{mgf}_X(t)\right|_{t=0} = \mathbb{E}[X^i].$$

An *analytic* function is one for which the Taylor series expansion converges to the function. In particular, if $\text{mgf}_X(t)$ is analytic around 0, then

$$\text{mgf}_X(t) = 1 + \mathbb{E}[X]t + \frac{\mathbb{E}[X^2]t^2}{2!} + \cdots$$

That earns this function the *moment generating function* name.

SHIFTING AND SCALING

Consider how shifting and scaling affects the moment generating function.

> **F79** If X has moment generating function $\text{mgf}_X(t)$, then for real a and b,
>
> $$\text{mgf}_{aX+b}(t) = \exp(tb)\,\text{mgf}_X(at).$$

Proof. Note that

$$\begin{aligned}\text{mgf}_{aX+b}(t) &= \mathbb{E}[\exp((aX+b)t)] \\ &= \mathbb{E}[\exp(bt)\exp(atX)] \\ &= \exp(bt)\,\text{mgf}_X(at).\end{aligned}$$

$\square$

Recall, that *standardizing* random variables means shifting down by the expected value and dividing by the standard deviation. This changes the moment generating function as follows.

> **F80** If X has mean μ and standard deviation σ, then for
>
> $$S = \frac{X - \mu}{\sigma},$$
>
> $\text{mgf}_S(t) = \exp(-t\mu/\sigma)\,\text{mgf}_X(t/\sigma).$

THE MGF AND THE SUM OF RANDOM VARIABLES

Suppose that $S_1, S_2, \ldots$ are random variables that have already been standardized so that they have mean 0 and standard deviation 1. So $\mathbb{E}[S_i] = 0$ and $\mathbb{E}[S_i^2] = \text{var}(S_i) = 1$.

Then if $\text{mgf}_{S_i}(t)$ is analytic, the first few terms are known, so

$$\text{mgf}_{S_i}(t) = 1 + \frac{t^2}{2} + \frac{\mathbb{E}[S_i^3]t^3}{3!} + \frac{\mathbb{E}[S_i^4]t^4}{4!} + \cdots.$$

Then

$$\mathbb{E}(S_1 + \cdots + S_n) = \mathbb{E}(S_1) + \cdots + \mathbb{E}(S_n) = 0,$$

and

$$\text{var}(S_1 + \cdots + S_n) = \text{var}(S_1) + \cdots + \text{var}(S_n) = n.$$

So then

$$W = \frac{S_1 + \cdots + S_n}{\sqrt{n}}$$

has mean 0 and standard deviation 1.

Then the moment generating function of W is

$$\begin{aligned}\text{mgf}_W(t) &= \text{mgf}_{S_1/\sqrt{n}}(t) \cdots \text{mgf}_{S_n/\sqrt{n}}(t) \\ &= \left(\text{mgf}_{S_i/\sqrt{n}}(t)\right)^n \\ &= \text{mgf}_{S_n}(t/\sqrt{n})^n\end{aligned}$$

In infinite series form:

$$\text{mgf}_W(t) = \left[1 + \frac{t^2}{2n} + \frac{\mathbb{E}[S_i^3]t^3}{3!n^{3/2}} + \frac{\mathbb{E}[S_i^4]t^4}{4!n^2} + \cdots\right]^n.$$

If $\text{mgf}_{S_i}(t)$ exists, then the limit of the right hand side as $n \to \infty$ is

$$\lim_{n \to \infty} \text{mgf}_W(t) = \exp(t^2/2).$$

THE NORMAL DISTRIBUTION

Is there a random variable whose mgf is $\exp(t^2/2)$? Yes! This distribution is called the normal distribution.

> **D83** Say that Z has a **standard normal distribution** if it has density
>
> $$f_Z(x) = \frac{1}{\sqrt{\tau}}\exp(-x^2/2).$$
>
> Write $Z \sim \mathsf{N}(0,1)$.

F81 If $Z \sim \mathsf{N}(0,1)$ then $\text{mgf}_Z(t) = \exp(t^2/2)$.

Proof. This can be calculated as

$$\text{mgf}_Z(t) = \mathbb{E}[\exp(tZ)]$$
$$= \int_{z \in \mathbb{R}} \exp(tz) \frac{1}{\sqrt{\tau}} \exp(-z^2/2) \, dz$$
$$= \int_{z \in \mathbb{R}} \frac{1}{\sqrt{\tau}} \exp(-(z-t)^2/2) \exp(t^2/2) \, dz.$$

Now substitute $s = z - t$ and $ds = dz$ to get

$$\text{mgf}_Z(t) = \exp(t^2/2) \int_{z \in \mathbb{R}} \frac{1}{\sqrt{\tau}} \exp(-z^2/2) \, ds$$

$$= \exp(t^2/2),$$

where the integral evaluates to 1 because it is just the integral over $\mathbb{R}$ of the density of a standard normal random variable. $\square$

This leads us to the Central Limit Theorem, which says that the standardized sum of random variables with finite variance approaches that of a normal random variable.

T10 The Central Limit Theorem
Suppose that $X_1, X_2, \ldots$ is an iid sequence of random variables with finite mean μ and variance σ^2. Then for all numbers a,

$$\lim_{n \to \infty} \mathbb{P}\left(\frac{X_1 + \cdots + X_n - n\mu}{\sigma\sqrt{n}} \leq a\right) = \mathbb{P}(Z \leq a),$$

where $Z \sim \mathsf{N}(0,1)$.

THE CDF OF A STANDARD NORMAL
Most advanced modern calculators can find $\mathbb{P}(Z \leq a)$ for any a. In R, the command `pnorm(a)` finds this value.

SOLVING THE STORY

Recall for a standard uniform, $U \sim \mathsf{Unif}([0,1])$, that $\mathbb{E}[U] = 1/2$ and $\text{var}(U) = 1/12$. So for

$$W = 4000U \sim \mathsf{Unif}([0, 4000]),$$

the mean is $\mathbb{E}[W] = (1/2)(4000) = 2000$ and the variance is $\text{var}(W) = 4000^2(1/12)$. Let $W_1, \ldots, W_5$ be iid W.

Armed with this information, one can standardize the sum inside the probability function using $W_1 + W_2 + \cdots + W_5 = S$:

$$p = \mathbb{P}(W_1 + \cdots + W_5 \geq 14000)$$
$$= \mathbb{P}(S - 5 \cdot 2000 \geq 14000 - 5 \cdot 2000)$$
$$= \mathbb{P}\left(\frac{S - 5 \cdot 2000}{\sqrt{5 \cdot 4000^2/12}} \geq \frac{14000 - 5 \cdot 2000}{\sqrt{5 \cdot 4000^2/12}}\right)$$
$$\approx \mathbb{P}(Z \geq 1.549193).$$

Using `1 - pnorm(1.549193)` in R gives $\boxed{p \approx 0.06066}$.

THE HALF INTEGER CORRECTION

For discrete random variables, the *half integer* correction can help the Central Limit Theorem be more accurate.

D84 When a random variable S must be an integer, for $i \in \mathbb{Z}$,

$$\mathbb{P}(S \leq i) = \mathbb{P}(S \leq i + 1/2),$$

and

$$\mathbb{P}(S \geq i) = \mathbb{P}(S \geq i - 1/2).$$

This is called the **half integer** correction.

E64 Suppose $D_1, \ldots, D_{10}$ are iid d6. Estimate $\mathbb{P}(D_1 + \cdots + D_{10} \leq 30)$ using the Central Limit Theorem.
Answer. For $D \sim$ d6, $\mathbb{E}[D] = (1+6)/2 = 3.5$ and $\text{var}(D) = 61.25$. Using the half-integer

correction and $D_1 + \cdots + D_{10} = S$ gives

$$p = \mathbb{P}(D_1 + \cdots + D_{10} \leq 30)$$
$$= \mathbb{P}(S \leq 30.5)$$
$$= \mathbb{P}(S - 10 \cdot 3.5 \leq 30.5 - 10 \cdot 3.5)$$
$$= \mathbb{P}\left(\frac{S - 10 \cdot 3.5}{\sqrt{10 \cdot 61.25}} \leq \frac{30.5 - 10 \cdot 3.5}{\sqrt{10 \cdot 5 \cdot 7/12}}\right)$$
$$\approx \mathbb{P}\left(Z \leq \frac{30.5 - 10 \cdot 3.5}{\sqrt{10 \cdot 5 \cdot 7/12}}\right)$$
$$= \mathbb{P}(Z \leq -0.8332\ldots).$$

This gives $\boxed{0.2023\ldots}$ as the estimate.

ENCOUNTERS

243. CLT FOR EXPONENTIALS

Suppose $T_1, T_2, \ldots, T_{18}$ are iid $\mathrm{Exp}(3.4)$. Estimate $\mathbb{P}(T_1 + \cdots + T_{18} \geq 6)$ using the Central Limit Theorem.

244. CLT FOR UNIFORMS

Suppose $A_1, A_2, \ldots, A_{18}$ are iid $\mathrm{Unif}([0, 2])$. Estimate $\mathbb{P}(A_1 + \cdots + A_{18} \geq 24.3)$ using the Central Limit Theorem.

245. CLT FOR GEOMETRICS

Suppose $T_1, T_2, \ldots, T_{18}$ are iid $\mathrm{Geo}(0.2)$. Estimate $\mathbb{P}(T_1 + \cdots + T_{18} \geq 100)$ using the Central Limit Theorem and the half-integer correction.

246. CLT FOR DICE

Suppose $D_1, D_2, \ldots, D_{18}$ are iid d8. Estimate $\mathbb{P}(D_1 + \cdots + D_{18} \leq 78)$ using the Central Limit Theorem and the half-integer correction.

247. ANOTHER EXPONENTIAL CLT

Suppose $T_1, \ldots, T_{10}$ are iid $\mathrm{Exp}(1.4)$. Estimate $\mathbb{P}(T_1 + \cdots + T_{10} \geq 9)$.

248. CLT FOR UNKNOWN DISTRIBUTIONS

If $S_1, \ldots, S_{15}$ have mean 15.2 and variance 24.3, then approximately what is $\mathbb{P}(S_1 + S_{15} > 240)$?

249. FACTORY ACCIDENTS

There are fifteen different subsections within a factory. Each subsection typically has a number of accidents (independently) that are modeled as uniform from 0 up to 20. Estimate with the Central Limit Theorem and the half-integer correction the probability that the number of accidents is at least 160.

250. CUSTOMER ARRIVALS

The time between customer arrivals is modeled as an iid sequence of random variables with density $4w^2 \exp(-2w)\mathbb{I}(w \geq 0)$. Estimate the probability that the time the fifth customer arrives is at least 8.2.

251. CHECKING THE CLT

Suppose $A_1, \ldots, A_{10}$ are iid $\mathrm{Exp}(2.1)$.
a. Estimate $\mathbb{P}(A_1 + \cdots + A_{10}) \geq 5$ using the CLT.
b. Find $\mathbb{P}(A_1 + \cdots + A_{10}) \geq 5$ using the fact that the sum of 10 iid exponentials is gamma distributed with parameters 10 and 2.1.

252. PRESCRIPTIONS

A pharmacy has 1000 customers, each of which has a 3% chance of picking up a prescription on a given day. Using the Central Limit Theorem, estimate the probability that at least 40 customers will pick up a prescription today.

Chapter 22: Normal Random Variables

THE RANGER LOOKED TOWARD the distant mountain. A simple model for the height of the mountain was normal with mean 10000 feet and standard deviation 2000 feet. What is the chance with this model that the mountain's height is at least 13000 feet?

Shifting and scaling standard normals

Standard normal random variables have mean 0 and variance 1. If these random variables are scaled and shifted, then the result is a another normal random variable, but with different mean and variance. The mean becomes the shift, and the variance is the square of the scaling.

D85 Let Z be a standard normal. For real numbers μ and σ, say that $\mu + \sigma Z$ is a **normal random variable** with mean μ and variance σ^2. Write $\mu + \sigma Z \sim \mathsf{N}(\mu, \sigma^2)$.

SCALING RANDOM VARIABLES
Other scaled random variables include exponentials and continuous uniforms over intervals. For A a standard exponential, A/λ is an exponential with rate λ. For U a standard uniform over $[0, 1]$, $a + bU \sim \mathsf{Unif}([a, a+b])$.

Recall that for $Z \sim \mathsf{N}(0, 1)$,
$$f(z) = \frac{1}{\sqrt{\tau}} \exp(-z^2/2)$$
$$\mathrm{mgf}_Z(t) = \exp(t^2/2).$$

The following then comes from the rules for densities and mgf obtained from shifted and scaled random variables.

F82 For $W \sim \mathsf{N}(\mu, \sigma^2)$, the density and mgf are as follows.
$$f_W(w) = \frac{1}{\sigma\sqrt{\tau}} \exp(-(x-\mu)^2/(2\sigma^2))$$
$$\mathrm{mgf}_W(t) = \exp(t\mu)\exp(t^2\sigma^2/2).$$

Solving the story

A standard normal can be multiplied by 2000 and added to 10000 to give a normal with mean 10000 and standard deviation 2000. Hence if H is the mountain height:

$$\mathbb{P}(H \geq 13000) = \mathbb{P}(10000 + 2000Z \geq 13000)$$
$$= \mathbb{P}(2000Z \geq 3000)$$
$$= \mathbb{P}(Z \geq 1.5).$$

Using `1 - pnorm(1.5)` in R gives an approximation for the probability of $\boxed{0.06680}$.

ADDING INDEPENDENT NORMALS

What happens when you add independent normals? The result is another normal random variable!

> **F83** Suppose $Z_1 \sim N(\mu_1, \sigma_1^2)$ and $Z_2 \sim N(\mu_2, \sigma_2^2)$ are independent. Then
> $$Z_1 + Z_2 \sim N(\mu_1 + \mu_2, \sigma_1^2 + \sigma_2^2).$$

Proof. Use the moment generating function to prove this. Recall that for independent random variables, the moment generating function of the sum equals the product of the mgfs.

$$\mathrm{mgf}_{Z_1+Z_2}(t) = \mathrm{mgf}_{Z_1}(t)\,\mathrm{mgf}_{Z_2}(t)$$
$$= \exp(t\mu_1)\exp(t^2\sigma_1^2/2))\exp(t\mu_2)\exp(t^2\sigma_2^2/2)$$
$$= \exp(t(\mu_1+\mu_2))\exp(t^2(\sigma_1^2+\sigma_2^2)/2).$$

This proves the result. □

This can be extended via induction to adding any number of independent normal random variables.

> **F84** If $(W_1, \ldots, W_n)$ are independent normal random variables where $W_i \sim N(\mu_i, \sigma_i^2)$, then
> $$W_1 + \cdots + W_n \sim N(\mu_1 + \cdots + \mu_n, \sigma_1^2 + \cdots + \sigma_n^2).$$

MULTIVARIATE NORMALS

If $(Z_1, Z_2, \ldots, Z_n)$ are iid standard normals, then the marginal distributions are standard normals.

By the results of the previous section, any linear combination of standard normals is also a normal. For instance,

$$W_1 = 3Z_1 - 4Z_2 + Z_3 + 4$$

will be normal. The parameters will be the mean and the variance.

$$W_1 \sim N(4, 3^2 + (-4)^2 + 1^2) \sim N(4, 26).$$

If we create n normals in this fashion from n iid standard normals, then the result is called a *multivariate normal*. Typically this is written using matrix notation as

$$W = AZ + \mu,$$

where $\mu = (\mu_1, \ldots, \mu_n)^T$ is a column vector of length μ, A is a real valued n by n matrix, and $Z = (Z_1, \ldots, Z_n)^T$ are iid standard normal random variables. For a_{ij} equal to the entry in the ith row and jth column of A, this means

$$W_i = \sum_{j=1}^{n} a_{ij} Z_j + \mu_i$$

COVARIANCE FOR THE MULTIVARIATE NORMAL

Consider
$$\mathrm{cov}(W_k, W_\ell)$$

Start with

$$\mathrm{cov}(W_k, W_\ell) = \mathrm{cov}\left(\sum_{i=1}^{n} a_{ki} Z_i, \sum_{j=1}^{n} a_{\ell j} Z_j\right),$$
$$= \sum_{i=1}^{n}\sum_{j=1}^{n} a_{ki} a_{\ell j} \, \mathrm{cov}(Z_i, Z_j).$$

Because for $i \neq j$, Z_i and Z_j are independent (and so uncorrelated), this sum reduces down to the single sum

$$\mathrm{cov}(W_k, W_\ell) = \sum_{i=1}^{n} a_{ki} a_{\ell i} \, \mathrm{cov}(Z_i, Z_i)$$
$$= \sum_{i=1}^{n} a_{ki} a_{\ell i},$$

because $\mathrm{cov}(Z_i, Z_i) = \mathrm{var}(Z_i) = 1$.

In matrix notation, this can be written simulataneously for all k and ℓ as
$$\Sigma = AA^T,$$
where A^T is the *transpose* of A defined as $A^T_{ji} = A_{ij}$. Here Σ is the n by n matrix whose (k, ℓ) entry is $\mathrm{cov}(W_k, W_\ell)$.

D86 If $X = (X_1, \ldots, X_n)$ is a multivariate real valued random variable, then the n by n matrix with $\Sigma_{ij} = \mathrm{cov}(X_i, X_j)$ is called the **covariance matrix** of X.

D87 If A is a real valued n by n matrix, and μ is a real valued n by 1 matrix, and $Z = (Z_1, \ldots, Z_n)^T$ consists of iid standard normal random variables, then
$$W = AZ + \mu,$$
is a multivariate normal random variable (aka **multinormal**) with mean μ and covariance matrix $\Sigma = AA^T$. Write
$$W \sim \mathsf{Multinormal}(\mu, \Sigma).$$

DENSITY OF A MULTIVARIATE NORMAL

The joint density of $Z_1, Z_2, \ldots, Z_n$ will be
$$\tau^{-n/2} \exp(-(z_1 + \cdots + z_n)^2/2),$$
and so for the density of the multivariate, it is necessary to understand how scaling and shifting changes joint densities.

The shifting part is easier. If $W = X + \mu$ for a multivariate distribution, then
$$f_W(w) = f_X(w - \mu).$$

For scaling in one dimension, recall that for $T = aS$,
$$f_T(t) = f_S(t/a)/\mathsf{a}.$$

The $|a|$ appears because of the way differential elements get stretched by a factor of $|a|$. To fit into a differential element of width $|dt|$, you must start with a differential element of width $|ds|/|a|$.

With multivariate scaling, a similar effect happens. For
$$Y = AX,$$
where X and Y are n dimensional vectors and A is an $n \times n$ matrix, the stretching of a differential element is by a factor called the *determinant* of A. To compensate for that, the density must be divided by the determinant of A. This leads to the following density for multivariate normal, presented without formal proof.

F85 For $W \sim \mathsf{Multinormal}(\mu, \Sigma)$, the density of W with n components is $f_W(w)$, which is
$$\frac{1}{\tau^{n/2}} \cdot \frac{1}{|\det(\Sigma)|^{1/2}} \exp\left(-\frac{(w-\mu)^T \Sigma^{-1} (w-\mu)}{2}\right)$$

ENCOUNTERS

253. SHIFTING AND SCALING NORMALS

Suppose $W \sim \mathsf{N}(34, 20)$. Then write $\mathbb{P}(W \leq 28)$ in terms of cdf_Z where $Z \sim \mathsf{N}(0, 1)$.

254. MORE SHIFTING AND SCALING NORMALS

If $R \sim \mathsf{N}(-4.2, 5.6)$, write $\mathbb{P}(R \geq -2)$ in terms of cdf_Z where Z is a standard normal random variable.

255. ADDING INDEPENDENT NORMALS

If $Z_1 \sim \mathsf{N}(2.4, 5.2)$ and $Z_2 \sim \mathsf{N}(-1.2, 5.2)$ are independent, what is the distribution of $Z_1 + Z_2$?

256. Adding Three Independent Normals

If $Z_1 \sim \mathsf{N}(1.9, 10.2)$, $Z_2 \sim \mathsf{N}(2.1, 13.2)$ and $Z_3 \sim \mathsf{N}(3.1, 11.2)$, what is the distribution of $Z_1 + Z_2 - Z_3$?

257. Multivariate Scaling and Shifting

Suppose that (Z_1, Z_2) are iid standard normal random variables, and

$$W_1 = 3Z_1 - Z_2 + 3$$
$$W_2 = -2Z_1 + Z_2 - 4.$$

What is the distribution of (W_1, W_2)?

258. Finding Multivariate Normal Parameters

Suppose that $X_1 = Z_1 + Z_2 + 4$ and $X_2 = Z_1 - Z_2 + 4$ where Z_1 and Z_2 are independent standard normal random variables. What is the distribution of (X_1, X_2)?

Chapter 23: Bayes' rule for densities

ALCHEMY IS TRICKY UNDER THE best of conditions. The Alchemist had developed a new recipe, but the same recipe (even when followed to the best of the Alchemist's ability) did not always result in a potent potion. The Alchemist decided to call the probability that a potion worked p, and originally (because of a lack of information) modeled p as $\mathsf{Unif}([0,1])$. The Alchemist then ran an experiment, making ten of the potions from the new recipe and trying each in turn. Given the result that there were four successes and six failures, what is the new distribution of p given this information?

Differentials and Bayes

Recall that the probability a random variable W falls into a differential interval dw is

$$\mathbb{P}(W \in dw) = f_W(w)\, dw,$$

where dw stands for both the differential interval and the length of the interval. The function $f_W(w)$ is the density of W. Alternatively, the density can be viewed as a type of derivative:

$$\frac{\mathbb{P}(W \in dw)}{dw} = f_W(w).$$

Conditioning on information about differential intervals is a bit different. Suppose that one is given the information that $Y \in dy$. That says that Y is arbitrarily close to y. Hence from a conditioning point of view, this is the same information as if $Y = y$. Consider using this idea with Bayes' rule.

If differentials could be cancelled just like regular variables, then the regular Bayes' rule gives

$$\begin{aligned}
f_{X|Y=y}(x)\, dx &= \mathbb{P}(X \in dx \mid Y \in dy) \\
&= \mathbb{P}(Y \in dy \mid X \in dx) \frac{\mathbb{P}(X \in dx)}{\mathbb{P}(Y \in dy)} \\
&= f_{Y|X=x}(y)\, dy \cdot \frac{f_X(x)\, dx}{f_Y(y)\, dy} \\
&= f_{Y|X=x}(y) \frac{f_X(x)}{f_Y(y)}\, dx,
\end{aligned}$$

and dividing both sides by dx gives

$$f_{X|Y=y}(x) = f_{Y|X=x}(y)\frac{f_X(x)}{f_Y(y)}.$$

Note that $f_{Y|X=x}(y)$ and $f_X(x)$ depends on x, while $f_Y(y)$ does not. This means that the normalizing constant $f_Y(y)$ can be found by integrating the other terms. This gives Bayes' rule for densities.

T11 Bayes' Rule for Densities

Suppose X and Y are real valued random variables with density f_X and f_Y respectively. Let y be a number such that $f_Y(y) > 0$. Then

$$f_{X|Y=y}(x) = \frac{f_{Y|X=x}(y) f_X(x)}{\int_w f_{Y|X=w}(y) f_X(w)\, dw}.$$

The density of X is called the *prior density*, while the density of the random variable X given the condition that $Y = y$, is called the *posterior density*.

Note that the denominator on the right hand side only depends on y after w is integrated out. And y is a constant on the left hand side of the equation. So another way to write Bayes' Rule for densities is as a *proportion*, using the $\propto$ symbol to hide the constant of proportionality.

Some examples of the $\propto$ symbol:
$$3x^2 \propto x^2$$
$$x^2 \propto 3x^2$$
$$6\sqrt{x} \propto \sqrt{x}$$

For Bayes's rule, the following holds.

F86 For random variables X and Y with densities f_X and f_Y, and y a number such that $f_Y(y) > 0$, it holds that
$$f_{X|Y=y}(x) \propto f_{Y|X=x}(y) f_X(x).$$

In words: the posterior density of X conditioned on Y is proportional to the prior density of X times the density of Y conditioned on X.

With this equation, you can then find the constant of proportionality by integrating C times the right hand side and for $x \in \mathbb{R}$ and then setting the result equal to 1.

Solving the story

In the Alchemist's story, p is a random variable that is uniform over $[0, 1]$. Hence it has density $f_p(a) = \mathbb{I}(a \in [0, 1])$. Let N denote the number of successful experiments for the 10 trials. Then the goal is to find
$$f_{p|N=4}(a) \propto f_{N|p=a}(4) f_p(a)$$

Here $[N \mid p] \sim \text{Bin}(10, p)$. So
$$f_{N|p=a}(4) = \binom{10}{4} a^4 (1-a)^{10-4} \propto a^4 (1-a)^6.$$

Multiplying by $f_p(a)$ gives
$$f_{N|p=a}(4) f_p(a) \propto a^4 (1-a)^6 \mathbb{I}(a \in [0,1]).$$

This is an *unnormalized density*. To normalize, integrate for all a:
$$\int_a a^4(1-a)^6 \mathbb{I}(a \in [0,1])\, da = \int_0^1 a^4(1-a)^6\, da$$
$$= 1/2310.$$

That last calculation comes from a neat little fact.

F87 For i and j integers,
$$\int_0^1 x^i (1-x)^j\, dx = \frac{i!j!}{(i+j+1)!}.$$

Dividing by $1/2310$ is the same as multiplying by 2310, so that makes the posterior density:
$$\boxed{f_{p|N=4}(a) = 2310 a^4 (1-a)^6 \mathbb{I}(a \in [0,1]).}$$

The Beta Distribution

The type of density that shows up in the posterior is called the *beta distribution*.

D88 Suppose X has density
$$f_X(x) \propto x^{a-1}(1-x)^{b-1} \mathbb{I}(x \in [0,1]),$$
where $a, b > 0$. Then say that X has the **beta distribution** with **parameters** a and b. Write $X \sim \text{Beta}(a, b)$.

> **Greek distributions**
> Like with the gamma distribution, it is customary to write out the English word Beta for the Greek Letter β. The Greek letter β (or at least the capitalized version B) is used for the function that gives the normalizing constant for the density.

When $a = b = 1$, the beta distribution is the same as a standard uniform over $[0, 1]$, so it can be thought of as a generalization of this distribution.

D89 The normalizing constant of the beta distribution is called the **beta function**, and is denoted by a capital Greek letter beta:

$$B(a,b) = \int_0^1 x^{a-1}(1-x)^{b-1}\,dx.$$

GREEK AND ROMAN LETTERS
The capital Greek letter beta and a capital Roman letter B look identical.

There is a simple relationship between the beta function and the gamma function.

F88 For $a, b > 0$,

$$B(a,b) = \frac{\Gamma(a)\Gamma(b)}{\Gamma(a+b)}.$$

BETA FUNCTIONS AND BINOMIAL COEFFICIENTS
This has the form of the inverse of a binomial coefficient. That is because the binomial coefficient multiplies the binomial density and $B(a,b)$ divides it.

ENCOUNTERS

259. ARCHYTAS MEDICAL

Archytas Medical Group believes a new drug has p chance of working, where p is a random variable with density $f_p(a) \sim \text{Beta}(2,3)$. They test the drug on five animals, three of whom show that the drug works. What is the posterior distribution on p given this information?

260. UPDATING RATE OF AN EXPONENTIAL

A modeler believes the prior distribution of λ is $\text{Exp}(0.01)$ and has a model of $[X \mid \lambda] = \text{Exp}(\lambda)$. Given $X = 4.3$, what is the posterior distribution of λ?

261. TOOTHPASTE TROUBLES

A factory produces 980 tubes of toothpaste in a day. The chance that any tube is defective is 0.001.

a. What is the chance that no tubes of toothpaste are defective?

b. What is the chance that at least two tubes of toothpaste are defective?

262. LARGE AND SMALL ORDERS

About 2% of the time, customers at a local restaurant have a large order than takes ten minutes or more. The restaurant has 40 customers in a typical lunch hour.

a. What is the chance that none of the customers is a large order?

b. What is the chance that there is at least one large order?

263. RONCO SURVEY GROUP

Ronco Survey Group knows that the chance of someone answering their phone and doing a survey is 4%. How many people do they have to call in order to make sure that the probability that they get at least 10 survey takers is at least 70

264. SOME ROCKET SCIENCE

An aerospace firm is developing several different varieties of rocket fuel mixtures. If each mixture has a 10% chance of meeting specifications, how many mixtures must they try in order to have at least an 80% chance that at least five mixtures are successful? (Feel free to use R in solving this problem.)

265. Dimer Medicine

Dimer Medicine creates 3 types of drugs for a particular illness. The first is effective in 50% of patients, the second in 37%, and the third in 5%. Let A denote the event that the drug is effective and $X \in \{1, 2, 3\}$ the drug that is given to the patient.

a. If a patient is equally likely to receive any of the three drugs, find the probability that both drug 1 is administered and it is effective.

b. If a patient is equally likely to receive any of the three drugs, what is the probability that the drug is effective on their illness? Hint: the event A is the disjoint union of three pieces.

$$A = (A \cap \{X=1\}) \cup (A \cap \{X=2\}) \cup (A \cap \{X=3\}).$$

c. If the drug is effective for the patient, what is the probability that the drug was of the third type?

266. Torrent Music

Torrent Music Co. signs artists that fall into high, middle, and low selling categories. High artists sell roughly 80% of the music, middle 15%, and low 5%. Only 2% of artists are high selling, 47% are middle selling, and 51% are low selling.

a. Given a sold piece of music, what is the chance that it belongs to a high selling artist?

b. What is the chance it belongs to a middle selling artist?

c. What is the chance it belongs to a low selling artist?

267. The assembly line

A consultant models the probability p that an item is defective on an assembly line as being uniform over the set $\{0, 0.01, 0.02, 0.03, 0.04, 0.05\}$. After testing 100 items that are believed to be independently defective or not defective, 4 are found to be defective. What is the distribution of p conditioned on this information?

268. Inspector Quinn

Inspector Quinn believes the chance of product failure is equally likely to be any of $\{0, 0.3, 0.5, 0.7, 1\}$. After testing 8 items, 3 are found to fail. What are the probabilities for the five different possibilities given this information?

269. Rush Hour

In a particular county during rush hour, 80% of cars contain one occupant, 10% contain 2, 5% contain 3, and 5% contains 4 or more.

Any car containing two or more occupants has a 90% chance of using the carpool lane, and 1% of cars containing only one occupant cheat and use the carpool lane.

Suppose a car is in the carpool lane. Given this information, what is the probability that the car contains 1, 2, 3, or 4+ occupants?

270. Uniform Bayes

Suppose that A is uniform from 1 to 4, and then B conditioned on A is uniform from 1 to A. Given $B = 2$, what is the distribution of A?

Chapter 24: The Multinomial Distribution

AGERLY THE ALCHEMIST watched the five potions on the table. For each potion, there were three outcomes. There was a 10% chance the potion would hurt the drinker, a 20% chance that it would do nothing, and a 70% chance the potion would help the drinker. If (N_1, N_2, N_3) represented the number of hurt, nothing, and help potions respectively, what would the covariance matrix of these random variables be?

More than two outcomes

In the Bernoulli process, each trial had only two outcomes, failure or success. What if more than two outcomes are allowed?

Suppose that there are m different outcomes possible. In the Story, there are three outcomes for the different types of potions. There were five potions being brewed. That means that 11231, 33212, and 33333 are all possible outcomes.

Altogether there are 3^5 different ways for the five potions to turn out. That can be simplified if instead of caring about exactly which potion was which type, the only question of interest is *how many* of each type of potion was created. In that case, if there were five trials, and the result was

represented numerically as

$$1, 1, 2, 3, 1$$

then the count of results would be the vector

$$(3, 1, 1),$$

indicating that there were 3 type 1 potions, 1 type 2 potion, and 1 type 3 potion. Similarly, 33212 gives the vector (1, 2, 2), and 33333 gives (0, 0, 5).

Let N_i denote the number of potions of type i created. Then because there were five potions to start $N_1 + N_2 + N_3 = 5$. Moreover, N_1, N_2, and N_3 are all binomially distributed. In particular, $N_1 \sim \text{Bin}(5, 0.1)$, since there are five potions (trials), with a 10% chance of being type 1 (success).

Similarly, $N_2 \sim \text{Bin}(5, 0.2)$ and $N_3 \sim \text{Bin}(5, 0.7)$. So these random variables are not identically distributed. They are also *not* independent, since again $N_1 + N_2 + N_3 = 5$. Knowing N_1 and N_2 allows the calculation of N_3, so they cannot all be independent.

The name given to the joint distribution of the N_i is *multinomial*.

> **D90** Suppose for $i \in \{1, \ldots, m\}$, $N_i \sim \text{Bin}(n, p_i)$, and $N_1 + \cdots + N_m = n$. Then say that $(N_1, \ldots, N_m)$ have a multinomial distribution with parameters $(n, p_1, p_2, \ldots, p_m)$. Write
>
> $$(N_1, \ldots, N_m) \sim \text{Multinomial}(n, p_1, \ldots, p_m).$$

Recall that the set of n-tuples where each coordinate is in $\{1, \ldots, m\}$ can be written as

$$\{1, \ldots, m\}^n.$$

So for instance, $1132 \in \{1, 2, 3\}^4$.

COUNTING OUTCOMES
For a set of outcomes of trials $(w_1, \ldots, w_n) \in \{1, \ldots, m\}^n$, let

$$f_{\text{count}}(w_1, \ldots, w_n) = \left(\sum \mathbb{I}(w_j = 1), \ldots, \sum \mathbb{I}(w_j = m)\right).$$

The next fact uses this notation to describe exactly what the multinomial random variables are counting.

F89 Suppose $W_1, \ldots, W_n$ are iid where $\mathbb{P}(W_j = i) = p_i$ for all $j \in \{1, \ldots, n\}$ and $i \in \{1, \ldots, m\}$, and $p_1 + \cdots + p_m = 1$. Then

$$f_{\text{count}}(W_1, \ldots, W_n) \sim \text{Multinomial}(n, p_1, \ldots, p_m).$$

Proof. Let $f_{\text{count}}(W_1, \ldots, W_n) = (N_1, \ldots, N_m)$.

For every i and j, consider the indicator random variable $\mathbb{I}(W_j = i)$. This is a Bernoulli distributed random variable with parameter p_i. Also, $\mathbb{I}(W_1 = i), \ldots, \mathbb{I}(W_n, i)$ are independent, and

$$N_i = \sum_{j=1}^n \mathbb{I}(W_j = i) \sim \text{Bin}(n, p_i)$$

because N_i is the sum of independent Bernoulli random variables.

In addition,

$$\sum_{i=1}^m \sum_{j=1}^n \mathbb{I}(W_j = i) = \sum_{j=1}^n \sum_{i=1}^m \mathbb{I}(W_j = i) = \sum_{j=1}^n 1 = n.$$

Hence the resulting distribution is multinomial. □

DENSITY OF THE MULTINOMIAL DISTRIBUTION

To understand the density of the multinomial distribution, first a bit of combinatorics must be introduced.

D91 Consider sequences of length n consisting of the numbers $\{1, \ldots, m\}$. The number of sequences where each number $i \in \{1, \ldots, m\}$ appears exactly n_i times is called the *multinomial coefficient*, and is written

$$\binom{n}{n_1, n_2, \ldots, n_m}$$

In terms of the f_{count} function, the multinomial coefficients count the number of inputs to the function that results in the count $n_1, \ldots, n_m$. That is,

$$\binom{n}{n_1, \ldots, n_m}$$

is equal to the number of $(w_1, \ldots, w_n)$ in $\{1, \ldots, m\}^n$ such that

$$f_{\text{count}}(w_1, \ldots, w_n) = (n_1, \ldots, n_m).$$

It is the multinomial coefficient because of the following fact about multinomials.

F90 For m and n positive integers, $(x_1 + \cdots + x_m)^n$ is equal to

$$\sum_{n_1 + n_2 + \cdots + n_m = n} \binom{n}{n_1, n_2, \ldots n_m} x_1^{n_1} x_2^{n_2} \cdots x_m^{n_m}.$$

This fact follows directly from the definition and the distributive law. For instance, consider $(x_1 + x_2 + x_3)^4$, which is equal to

$$(x_1 + x_2 + x_3)(x_1 + x_2 + x_3)(x_1 + x_2 + x_3)(x_1 + x_2 + x_3)$$

To multiply this out, choose one term from the first factor, one from the second, one from the third, and finally one from the fourth. This is exactly like our random variables $W_1, \ldots, W_4$, where $W_i \in \{1, 2, 3\}$ for all i.

For instance, if the choice of terms from each factor was x_1, x_2, x_1, x_2, then multiplying gives $x_1^2 x_2^2$. Because multiplication of the x_i is commutative, the number of ways of getting $x_1^2 x_2^2$ equals the number of ways of choosing the first term twice, the second term twice, and the third term zero times from among the four products.

Now use these coefficients to write down the density of multinomial random variables.

> **F91** For $(N_1, \ldots, N_m) \sim \text{Multinomial}(n, p_1, \ldots, p_m)$, the density
> $$f_{(N_1,\ldots,N_m)}(n_1, \ldots, n_m)$$
> equals
> $$\binom{n}{n_1, n_2, \ldots, n_m} p_1^{n_1} p_2^{n_2} \cdots p_m^{n_m}.$$

Proof. Consider an outcome of the trials $(w_1, \ldots, w_n) \in \{1, \ldots, m\}^n$ where
$$f_{\text{count}}(w_1, \ldots, w_n) = (n_1, \ldots, n_m).$$

In our problem $\mathbb{P}(W_i = w_i) = p_{w_i}$. For all $p_j > 0$, there is also a useful product form of this expression:
$$\mathbb{P}(W_i = w_i) = \prod_{j=1}^m p_j^{\mathbb{I}(w_i=j)}.$$

When $w_i = j$, $\mathbb{I}(w_i = j) = 1$ and $p_j^{\mathbb{I}(w_i=j)} = p_j$. when $w_i \neq j$, $p_j^{\mathbb{I}(w_i=j)} = p_j^0 = 1$. Taking the product over all j picks out p_{w_i}.

Then
$$\begin{aligned}
p &= \mathbb{P}((W_1, \ldots, W_n) = (w_1, \ldots, w_n)) \\
&= \prod_{i=1}^n \mathbb{P}(W_i = w_i) \\
&= \prod_{i=1}^n \prod_{j=1}^m p_j^{\mathbb{I}(w_i=j)} \\
&= \prod_{j=1}^m \prod_{i=1}^n p_j^{\mathbb{I}(w_i=j)} \\
&= \prod_{j=1}^m p_j^{\sum_{i=1}^n \mathbb{I}(w_i=j)} \\
&= \prod_{j=1}^m p_j^{n_j}
\end{aligned}$$

So then the question is, *how many* $(w_1, \ldots, w_n)$ are there such that $(N_1, \ldots, N_m) = (n_1, \ldots, n_m)$. And that is exactly what the multinomial coefficient gives. Let A be the set of $(w_1, \ldots, w_n)$ such that
$$f_{\text{count}}(w_1, \ldots, w_n) = (n_1, \ldots, n_m).$$

Then
$$\#(A) = \binom{n}{n_1, n_2, \ldots, n_m}.$$

Moreover,
$$\begin{aligned}
p &= \mathbb{P}((N_1, \ldots, N_m) = (n_1, \ldots, n_m)) \\
&= \sum_{(w_1,\ldots,w_n) \in A} \mathbb{P}((W_1, \ldots, W_n) = (w_1, \ldots, w_n)) \\
&= \sum_{(w_1,\ldots,w_n) \in A} \prod_{j=1}^m p_j^{n_j} \\
&= \binom{n}{n_1, n_2, \ldots, n_m} \prod_{j=1}^m p_j^{n_j}.
\end{aligned}$$

$\square$

So how are multinomial coefficients actually calculated? Use the following formula.

F92 For $n_1, \ldots, n_m$ nonnegative integers that sum to n,
$$\binom{n}{n_1, n_2, \ldots, n_m} = \frac{n!}{n_1! n_2! \cdots n_m!}.$$

The proof is a bit tricky to write formally, so here just an informal derivation will be given.

Think of the multinomial as an m stage experiment. At the first stage, n_1 spots are chosen from the n available to have $W_j = 1$. So that can be done in
$$\binom{n}{n_1} = \frac{n!}{n_1!(n-n_1)!}$$
ways. At the second stage, choose n_2 spots from the remaining $n - n_1$ spots. This can be done in
$$\binom{n-n_1}{n_2} = frac(n-n_1)!n_2!(n-n_1)!$$
ways.

The number of ways to choose in the first two stages of the experiment are then
$$\binom{n}{n_1}\binom{n-n_1}{n_2} = \frac{n!}{n_1!(n-n_1)!}\frac{(n-n_1)!}{n_2!(n-n_1-n_2)!},$$
and two of the factorials cancel to give
$$\binom{n}{n_1}\binom{n-n_1}{n_2} = \frac{n!}{n_1!n_2!(n-n_1-n_2)!},$$

Then multiplying by
$$\binom{n-n_1-n_2}{n_3}$$
gives
$$\frac{n!}{n_1!n_2!n_3!(n-n_1-n_2-n_3)!}.$$

Carrying this through to all the stages of the experiment gives (after cancellation)
$$\frac{n!}{n_1!n_2!\cdots n_m!(n-n_1-\cdots-n_m)!},$$
however, $n-n_1-n_2-\cdots-n_m = 0$, and $0! = 1$, so it disappears from the product.

CHAPTER 24: THE MULTINOMIAL DISTRIBUTION

E65 If $(N_1, N_2, N_3) \sim$ Multinomial$(5, 0.1, 0.2, 0.7)$, then what is the chance that $(N_1, N_2, N_3) = (0, 2, 3)$?
Answer. Using our formula, this is
$$\frac{5!}{0!2!3!}(0.1)^0(0.2)^2(0.7)^3,$$
or $\boxed{0.1372}$.

MOMENTS OF MULTINOMIAL RANDOM VARIABLES

The mean of each component of the multinomial follows directly from its marginal distribution, which is binomial. Recall that the mean of a binomial is the product of the two parameters.

F93 For $(N_1, \ldots, N_m) \sim$ Multinomial$(n, p_1, \ldots, p_m)$,
$$\mathbb{E}[(N_1, \ldots, N_m)] = (np_1, np_2, \ldots, np_m).$$

A similar situation holds for variance.

F94 For $(N_1, \ldots, N_m) \sim$ Multinomial$(n, p_1, \ldots, p_m)$, for each $i \in \{1, \ldots, m\}$,
$$\text{var}(N_i) = np_i(1-p_i).$$

COVARIANCE OF MULTINOMIAL

Covariance is a bit trickier, since the components of the multinomial are *not* independent. If X_i is high, that leaves fewer spots to be taken by type j, so X_j

should be lower on average. How low is given by the following fact.

F95 For $i \neq j$, $\text{cov}(X_i, X_j) = -np_i p_j$.

Proof. Recall that

$$\text{cov}(X_i, X_j) = \mathbb{E}[X_i X_j] - \mathbb{E}[X_i]\mathbb{E}[X_j]$$
$$= \mathbb{E}[X_i X_j] - np_i np_j.$$

Say X_i is known. That gives $n - X_i$ spots for X_j. Moreover, because these spots are conditioned to not draw $W_k = i$, there is a $p_j/(1 - p_i)$ chance that $W_k = j$.

Use the Fundamental Theorem of Probability to say

$$\mathbb{E}[X_i X_j] = \mathbb{E}[\mathbb{E}[X_i X_j \mid X_i]]$$
$$= \mathbb{E}[X_i \mathbb{E}[X_j \mid X_i]]$$
$$= \mathbb{E}[X_i(n - X_i)p_j/(1 - p_i)]$$
$$= \frac{p_j}{1 - p_i}\mathbb{E}[nX_i - X_i^2]$$
$$= \frac{p_j}{1 - p_i}[n^2 p_i - \text{var}(X_i) - \mathbb{E}[X_i]^2]$$
$$= \frac{p_j}{1 - p_i}[n^2 p_i - np_i(1 - p_i) - [np_i]^2]$$
$$= \frac{p_j}{1 - p_i}[n^2 p_i(1 - p_i) - np_i(1 - p_i)]$$
$$= -np_i p_j + n^2 p_i p_j.$$

Hence

$$\text{cov}(X_i, X_j) = -np_i p_j + n^2 p_i p_j - n^2 p_i p_j,$$

and simplifying finishes the proof. □

Solving the Alchemist's Story

In the story, the probabilities of the three outcomes were 10%, 20%, and 70%, and the number of trials was 5, making the covariance matrix for the outcomes

$$\begin{pmatrix} 0.05 & -0.10 & -0.35 \\ -0.10 & 0.20 & -0.70 \\ -0.35 & -0.70 & 2.45 \end{pmatrix}$$

Encounters

271. Probabilities for Multinomials

If $(X_1, X_2, X_3, X_4) \sim$ Multinomial$(10, 0.3, 0.5, 0.1, 0.1)$, what is the chance that (X_1, X_2, X_3, X_4) equals $(3, 5, 1, 1)$?

272. More Probabilities of Multinomials

If $(A_1, A_2, A_3) \sim$ Multinomial$(7, 0.3, 0.3, 0.4)$, what is the chance that $(A_1, A_2, A_3) = (4, 2, 1)$?

273. Means of Multinomials

If $(X_1, X_2, X_3, X_4) \sim$ Multinomial$(10, 0.3, 0.5, 0.1, 0.1)$, what is $\mathbb{E}[(X_1, X_2, X_3, X_4)]$?

274. Multinomials Means

If $(A_1, A_2, A_3) \sim$ Multinomial$(7, 0.3, 0.3, 0.4)$, what is $\mathbb{E}[(A_1, A_2, A_3)]$?

275. Multinomial Covariance

If $(X_1, X_2, X_3, X_4) \sim$ Multinomial$(10, 0.3, 0.5, 0.1, 0.1)$, what is the covariance matrix for (X_1, X_2, X_3, X_4)?

276. More Covariances of Multinomials

If $(A_1, A_2, A_3) \sim$ Multinomial$(7, 0.3, 0.3, 0.4)$, what is the covariance matrix for (A_1, A_2, A_3)?

Chapter 25: Tail Inequalities

For the coming battle, the King expected the Vassal to send on average 100 archers to bolster his army. The King wondered, then, given the average to be 100, what was the largest the chance could be that the Vassal would send at least 300 archers?

Expectation as Balance Point

One way to view the expected value of a random variable is to think about carving the density out of wood. Then physically, the value $\mathbb{E}(X)$ is the point where if a fulcrum was placed underneath the density, it would balance exactly and not tip either to the right or the left. See the figure below.

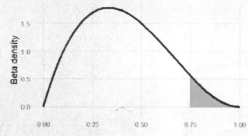

A density in blue with a red fulcrum and green tail

The shaded region over $[0.75, 1]$ in the figure above has area equal to $\mathbb{P}(X \geq 0.75)$. An event of the form $(X \geq a)$ for some a is called a *tail* (or more specifically a *right tail* or *upper tail*) event.

Given that the fulcrum (expected value) for the density in the figure is 0.4, the area of the right tail cannot be too large, or the figure would tip over when made in real life.

This idea, that the right tail cannot be too large given the expected value of a random variable, goes by the name *Markov's Inequality*.

> **T12** Markov's Inequality
> For an integrable random variable X, and $a > 0$,
> $$\mathbb{P}(|X| \geq a) \leq \frac{\mathbb{E}[|X|]}{a}.$$

Note that the theorem deals with a nonnegative random variable $|X|$. This is because if the right tail can be balanced with a left tail that is negative, the expected value does not prevent the right tail from being arbitrarily close to 1. But if the random variable cannot fall below 0, then the inequality holds.

The proof is fairly simple. The key idea is that for any positive numbers x and a
$$x\mathbb{I}(x \geq a) \geq a\mathbb{I}(x \geq a).$$

To see why this is true, first suppose $x < a$. In this case both sides are 0. If $x \geq a$ then the inequality reduces to $x\mathbb{I}(x \geq a) = x \geq a = a\mathbb{I}(x \geq a)$.

Also, since both x and $\mathbb{I}(x < a)$ are at least 0,
$$x\mathbb{I}(x < a) \geq 0.$$

With these inequalities in mind, the proof of Markov's inequality is as follows.

Proof. Let $a \geq 0$. Then
$$|X| = |X|\mathbb{I}(|X| \geq a) + |X|\mathbb{I}(|X| < a)$$
$$\geq a\mathbb{I}(|X| \geq a).$$

Taking the expected value of both sides gives

$$\mathbb{E}[|X|] \geq a\mathbb{E}[\mathbb{I}(|X| \geq a)] = a\mathbb{P}(|X| \geq a),$$

and rearranging gives the result. □

Solving the Story

In the Story, let A denote the number of archers sent by the Vassal. The King knows that $\mathbb{E}[A] = 100$, and wants an upper bound on $\mathbb{P}(A \geq 300)$. Since $A \geq 0$, $|A| = A$. From Markov's inequality, this is

$$\mathbb{P}(A \geq 300) \leq \frac{\mathbb{E}(A)}{300} = \frac{100}{300} = \frac{1}{3}.$$

So the King upper bounds the probability by $\boxed{0.3334}$.

> **Rounding upper bounds**
> For this problem, the last digit of the probability was rounded up. This is because the goal was to find an *upper bound* on the value. If the goal had been to find a *lower bound*, then the value would have been rounded down.

Chebyshev's inequality

Markov's inequality seems so simple, but is the basis of many other important tail inequalities in probability. One such is now known as *Chebyshev's inequality*.

> **Teacher and Student**
> Pafnuty Chebyshev was actually Andrey Markov's teacher. Chebyshev was the first to create what is now called Markov's inequality, so sometimes Markov's inequality is called the first Chebyshev inequality and Chebyshev's inequality is the second Chebyshev inequality.

This inequality gives an upper bound which is usually (although not always) better than the straight Markov's inequality. Moreover, it gives a bound on the sum of the upper and lower tails. And it works on random variables that are both positive and negative.

The idea is to apply Markov's inequality to the new random variable $(X - \mathbb{E}(X))^2$. Recall that the mean of this variable is the variance of X.

> **F96** Chebyshev's inequality
> For X an integrable random variable with finite standard deviation, and $a > 0$,
> $$\mathbb{P}(|X - \mathbb{E}[X]| \geq a) \leq \frac{\text{var}(X)}{a^2}.$$

Proof. By Markov's inequality,

$$\mathbb{P}((X - \mathbb{E}[X])^2 \geq a^2) \leq \frac{\mathbb{E}[(X - \mathbb{E}[X])^2]}{a^2}$$
$$= \frac{\text{var}(X)}{a^2}.$$

Then using

$$((X - \mathbb{E}[X])^2 \geq a^2) \leftrightarrow (|X - \mathbb{E}[X]| \geq a)$$

finishes the proof. □

> **E66** Suppose $\mathbb{E}[T] = 15$ and $\text{var}(T) = 25$. Upper bound $\mathbb{P}(T \geq 25)$ using Chebyshev's inequality.
> *Answer.* Note that $(a \geq b) \to (|a| \geq b)$, so
> $$(T \geq 25) = (T - 15 \geq 10) \to (|T - 15| \geq 10),$$
> so
> $$\mathbb{P}(T \geq 25) \leq \mathbb{P}(|T - 15| \geq 10)$$
> $$\leq \frac{\text{var}(T)}{10^2}$$
> $$= 25/100,$$
> so $\boxed{25\%}$ is an upper bound on the tail probability.

You can also phrase Chebyshev's inequality in terms of how many standard deviations away from the mean the random variable is.

F97 Suppose X has a finite standard deviation. Then for $k \geq 1$,
$$\mathbb{P}(|X - \mathbb{E}[X]| \geq k\,\mathrm{sd}(X)) \leq 1/k^2.$$

Proof. Plug $a = k\,\mathrm{sd}(X)$ into Chebyshev's inequality. □

SAMPLE AVERAGES

Recall that the sample average of the first n of $X_1, X_2, \ldots$ iid X where X is an integrable random variable is
$$S_n = \frac{X_1 + \cdots + X_n}{n}.$$

The Strong Law of Large Numbers indicates that with probability 1, $S_n \to \mathbb{E}[X]$. But it does not tell us how quickly the convergence occurs.

Unfortunately, Markov's inequality does not improve for sample averages as n increases. $\mathbb{E}[S_n] = \mathbb{E}[X]$, so
$$\mathbb{P}(S_n \geq a) \leq \mathbb{E}[S_n]/a = \mathbb{E}[X]/a.$$

On the other hand, if $\mathrm{var}(X) < \infty$, then Chebyshev's inequality does improve with n.

F98 Let $X_1, X_2, \ldots$ iid X with $\mathbb{E}[X] = \mu$, $\mathrm{sd}(X) = \sigma$, $S_n = (X_1 + \cdots + X_n)/n$, then
$$\mathbb{P}\left(\left|\frac{S_n - \mu}{\sigma}\right| \geq k\right) \leq \frac{1}{nk^2}.$$

Proof. Recall
$$\mathrm{var}(S_n) = \frac{\mathrm{var}(X_1 + \cdots + X_n)}{n^2}$$
$$= \frac{\mathrm{var}(X_1) + \cdots + \mathrm{var}(X_n)}{n^2}$$
$$= \frac{n\,\mathrm{var}(X_1)}{n^2}$$
$$= \frac{\mathrm{var}(X_1)}{n}.$$

Hence
$$\mathbb{P}\left(\left|\frac{S_n - \mu}{\sigma}\right| \geq k\right) = \mathbb{P}(|S_n - \mu| \geq k\sigma)$$
$$\leq \frac{\sigma^2}{n(k\sigma)^2}$$
$$= \frac{1}{nk^2}.$$
□

E67 Let $A_1, A_2, \ldots$ be iid A, and $S_n = (A_1 + \cdots + A_n)/n$. For A with mean 3.2 and standard deviation 1.6, how large does n have to be before $\mathbb{P}((A_1 + \cdots + A_n)/n > 4) \leq 0.1$ (using Chebyshev)?

Answer. Let $S_n = (A_1 + \cdots + A_n)/n$. First let's standardize S_n recalling that it has mean 3.2 and standard deviation $1.6/\mathrm{sqrt}(n)$. Then
$$\mathbb{P}(S_n > 4) = \mathbb{P}\left(\left|\frac{S_n - 3.2}{1.6/\mathrm{sqrt}(n)}\right| > \frac{4 - 3.2}{1.6/\mathrm{sqrt}(n)}\right)$$
$$= \mathbb{P}\left(\left|\frac{S_n - 3.2}{1.6/\mathrm{sqrt}(n)}\right| > \frac{\mathrm{sqrt}(n)}{2}\right)$$

At this point, invoke Chebyshev's inequality to say
$$\mathbb{P}(S_n > 4) \leq \left(\frac{2}{\mathrm{sqrt}(n)}\right)^2 = \frac{4}{n}.$$

Setting $4/n \leq 0.1$ gives $\boxed{n \geq 40}$.

RELATIVE ERROR

Now, usually the goal of using a sample average is to estimate the mean value of a random variable. Often the *relative error* is required.

D92 The **relative error** in estimate $\hat{a}$ for a is
$$\left|\frac{\hat{a}}{a} - 1\right|$$

For instance, if the true answer is 3, and the estimate is 3.3, then there is $|(3.3/3) - 1| = 10\%$ relative error in the estimate.

Now consider bounding the probability that the relative error in the sample average as an estimate of the mean is larger than ϵ. Using $|ab| = |a| \cdot |b|$ gives

$$\mathbb{P}\left(\left|\frac{S_n}{\mu} - 1\right| \geq \epsilon\right) = \mathbb{P}\left(|S_n - \mu| \geq \epsilon|\mu|\right)$$
$$\leq \frac{\sigma^2/n}{[\epsilon|\mu|]^2}$$
$$= \frac{1}{n} \cdot \epsilon^{-2} \frac{\sigma^2}{\mu^2}.$$

This last factor σ^2/μ^2 is unitless, and is given the name *relative variance*. The square root of the relative variance is called the *relative standard deviation* or *coefficient of variation*.

> **D93** For a random variable with nonzero mean and finite variance, the relative variance is
>
> $$\text{var}_{\text{rel}}(X) = \frac{\text{var}(X)}{\mathbb{E}[X]^2} = \frac{\mathbb{E}[X^2]}{\mathbb{E}[X]^2} - 1.$$
>
> The relative standard deviation") also known as 'r term("coefficient of variation is
>
> $$\text{sd}_{\text{rel}}(X) = \frac{\text{sd}(X)}{|\mathbb{E}[X]|} = \sqrt{\text{var}_{\text{rel}}(X)}.$$

So Chebyshev's inequality says that to get a nontrivial bound on the probability that the relative error of the sample deviation is large, the number of samples must be inversely proportional to the square of the relative error and proportional to the relative variance. That is,

$$n = \Theta\left(\epsilon^{-2} \cdot \frac{\text{var}(X)}{\mathbb{E}[X]^2}\right)$$

A look at how the Central Limit Theorem behaves gives a similar result.

This idea is often called the *Monte Carlo error bound*, and represents a hard limit on how well sample averages can estimate their mean.

ENCOUNTERS

277. MARKOV'S INEQUALITY

Suppose $T \geq 0$ has $\mathbb{E}[T] = 2.3$. Upper bound $\mathbb{P}(T \geq 6)$.

278. MORE MARKOV'S INEQUALITY

If $\mathbb{E}[X] \leq 10$, give an upper bound on $\mathbb{P}(X \geq 50)$.

279. CHEBYSHEV'S INEQUALITY

Suppose R has mean 12.2 and standard deviation 4.3. Use Chebyshev's inequality to bound $\mathbb{P}(R \leq 4)$.

280. MORE CHEBYSHEV'S INEQUALITY

Suppose A has mean 8.6 and standard deviation 1.2.

a. Use Markov's inequality to upper bound $\mathbb{P}(A \geq 13)$.

b. Use Chebyshev's inequality to bound $\mathbb{P}(A \geq 13)$.

281. CHEBYSHEV VIA STANDARD DEVIATION

What is the largest chance possible that a random variable is at least 3 standard deviations away from its mean?

282. MORE CHEBYSHEV VIA STANDARD DEVIATION

What is the largest chance possible that a random variable is at least 2.3 standard deviations away from its mean?

CHAPTER 25: TAIL INEQUALITIES

283. SAMPLE AVERAGES AND CHEBYSHEV

Suppose $N_1, N_2, \ldots$ are iid N, where $\mathbb{E}[N] = 3.2$ and $\text{sd}(N) = 6.2$. Let $S_n = (N_1 + \cdots + N_n)/n$. How large must n be in order for Chebyshev to upper bound $\mathbb{P}(S_n < 0)$ by 0.2?

284. MORE SAMPLE AVERAGES AND CHEBYSHEV

Suppose that $T_1, T_2, \ldots$ are iid $\text{Exp}(2.1)$. How large must n be for Chebyshev to guarantee that

$$\mathbb{P}\left(\frac{T_1 + \cdots + T_n}{n} > 0.5\right) \leq 0.01?$$

285. CONSTRUCTION WOES

A building is believed to require 0.8 years on average to complete, with a standard deviation of 0.5 years. Give the best bound (either Markov or Chebyshev) for the following.
a. The building takes at least a year to build.
b. The building takes at least two years to build.

286. THE VACCINE

A vaccine is expected to be ready on average in 13 months, with a standard deviation of 2 months. Give an upper bound on the chance that the vaccine is not ready 16 months from now.

Chapter 26: Chernoff Inequalities

HOURGLASSES WERE ALL THE rage at the market this year. The Merchant modeled the demand among the nobility as $D \sim \text{Pois}(112.2)$. How can $\mathbb{P}(D \geq 150)$ be upper bounded using the moment generating function of D?

The Chernoff Inequality

Recall that our tail inequalities begin with Markov's inequality for $a > 0$:

$$\mathbb{P}(|X| \geq a) \leq \frac{\mathbb{E}[|X|]}{a}.$$

By applying Markov's inequality to $(X - \mathbb{E}[X])^2$, Chebyshev's inequality is obtained:

$$\mathbb{P}(|X - \mathbb{E}[X]| \geq a) \leq \frac{\text{var}(X)}{a^2}.$$

Markov's inequality assumes only that X is integrable and Chebyshev's inequality assumes that X has a finite standard deviation. A Chernoff inequality makes the greatest assumption: here X must have a moment generating function that is finite for some $t \neq 0$.

F99 Chernoff's inequality
For $a \in \mathbb{R}$, and random variable X, for any $t > 0$,
$$\mathbb{P}(X \geq a) \leq \frac{\text{mgf}_X(t)}{\exp(ta)},$$
and for $t < 0$,
$$\mathbb{P}(X \leq a) \leq \frac{\text{mgf}_X(t)}{\exp(ta)},$$

Proof. For $t > 0$, $x \mapsto \exp(tx)$ is an increasing function. Hence

$$\mathbb{P}(X \geq a) = \mathbb{P}(\exp(tX) \geq \exp(ta)) \leq \frac{\mathbb{E}[\exp(tX)]}{\exp(ta)}$$

by Markov's inequality. Similarly, using $t < 0$, the map $x \mapsto \exp(tx)$ is a decreasing function, making

$$\mathbb{P}(X \leq a) = \mathbb{P}(\exp(tX) \geq \exp(ta))$$
$$\leq \frac{\mathbb{E}[\exp(tX)]}{\exp(ta)}.$$

□

Rubin's inequality?
Chernoff's inequality continues our tradition of not naming tail inequalities after their inventors. This particular inequality was actually first created by Herman Rubin but was popularized by Chernoff's work.

Unlike Markov and Chebyshev inequalities, Chernoff's inequalities actually contain an infinite number of choices! They hold for any t where the moment generating function is finite, so to get the best inequality, it is necessary to choose the value of t that gives the best result.

Solving the Story

To solve the story with Chernoff's inequality, the moment generating function of a Poisson is needed.

> **F100** For $X \sim \text{Pois}(\mu)$,
> $$\text{mgf}_X(t) = \exp(\mu(\exp(t) - 1)).$$

Proof. The calculation of the mgf can be done as follows. First recall the density of a Poisson random variable.
$$f_X(i) = \exp(-\mu)\frac{\mu^i}{i!}\mathbb{I}(i \in \{0, 1, 2, \ldots\}).$$

For any number t,
$$\text{mgf}_X(t) = \mathbb{E}[\exp(tX)]$$
$$= \int_i \exp(ti) f_X(i)\, d\#$$
$$= \sum_{i \in \{0,1,2,\ldots\}} \exp(ti) \exp(-\mu)\frac{\mu^i}{i!}$$
$$= \exp(-\mu) \sum_{i \in \{0,1,2,\ldots\}} \exp(ti)\frac{\mu^i}{i!}$$
$$= \exp(-\mu) \sum_{i \in \{0,1,2,\ldots\}} \frac{[\exp(t)\mu]^i}{i!}$$
$$= \exp(-\mu)\exp(\exp(t)\mu)$$
$$= \exp(\mu(\exp(t) - 1)).$$
$\square$

So Chernoff's inequality gives that
$$\mathbb{P}(X \geq a) \leq f(t, \mu, a) = \frac{\exp(\mu(\exp(t) - 1))}{\exp(ta)}.$$

To get the best bound, choose t to minimize $f(t, \mu, a)$. Since $f > 0$, use the following trick:
$$\arg\max_t f(t, \mu, a) = \arg\max_t \ln(f(t, \mu, a)).$$

So
$$\arg\max_t f(t, \mu, a) = \arg\max_t \mu(\exp(t) - 1) - ta.$$

The derivative with respect to t gives
$$[\mu(\exp(t) - 1) - ta]' = \mu\exp(t) - a,$$

which is positive if $\exp(t) > a/\mu$ and negative for $\exp(t) < a/\mu$. Hence
$$\arg\max_t f(t, \mu, a) = \ln(a/\mu).$$

For the story of the Merchant, $a = 140$, $\mu = 112.2$, $\exp(t) = 1.247772$, and
$$\mathbb{P}(D \geq 140) \leq \frac{\exp(112.2(1.247772 - 1))}{1.247772^{140}},$$

which is at most $\boxed{0.04116}$.

This is much better bound that either Markov's inequality which gives $112.2/140 = 0.8014\ldots$ or Chebyshev's inequality, which gives $112.2/(140 - 112.2)^2 = 0.145\ldots$.

Now that the t has been determined, it can be used without rederiving it each time for the Poisson.

> **F101** For $X \geq \text{Pois}(\mu)$, both $\mathbb{P}(X \geq a)$ and $\mathbb{P}(X \leq a)$ are at most
> $$\frac{\exp(a - \mu)}{(a/\mu)^a}.$$

Sums of Random Variables

Chernoff's bound works especially well with sums of random variables. Recall that if $S = S_1 + \cdots + S_n$, where the S_i are iid, then
$$\text{mgf}_S(t) = \text{mgf}_{S_1}(t)\text{mgf}_{S_2}(t)\cdots\text{mgf}_{S_n}(t)$$
$$= [\text{mgf}_{S_1}(t)]^n.$$

Hence for $t > 0$,
$$\mathbb{P}(S \geq a) \leq \text{mgf}_{S_1}(t)^n \exp(-ta),$$

and for $t < 0$,
$$\mathbb{P}(S \leq a) \leq \text{mgf}_{S_1}(t)^n \exp(-ta),$$

> **E68** For $X \sim \text{Bin}(5, 0.2)$, use Chernoff's

bound to bound $\mathbb{P}(X \geq 2)$.

Answer. Recall that a binomial random variable is the sum of iid Bernoulli random variables. That is, $X = B_1 + \cdots + B_n$ where $B_i \sim \text{Bern}(p)$. Hence $\text{mgf}_B(t) = \text{mgf}_{B_1}(t)^n$, and

$$\text{mgf}_{B_i}(t) = e^0(1-p) + e^t(p) = 1 - p(1-e^t).$$

Hence

$$\mathbb{P}(X \geq 2) \leq \frac{[1 - 0.2(1-e^t)]^5}{\exp(2t)}$$

$$= \left(\frac{1 - 0.2(1-e^t)}{\exp((2/5)t)}\right)^5$$

$$= (0.8\exp(-(2/5)t) + 0.2\exp((3/5)t))^5$$

$$= g(s)^5,$$

where $s = e^t$ and $g(s) = 0.8s^{-2/5} + 0.2s^{3/5}$.

Since $g(s) \geq 0$, both $g(s)$ and $g(s)^5$ have the same argument maximum. What value of s minimizes $g(s)$?

$$[0.8s^{-2/5} + 0.2s^{3/5}]' = -0.32s^{-7/5} + 0.12s^{-2/5}$$
$$= s^{-7/5}[-0.32 + 0.12s].$$

That means that the derivative is negative when $s < 0.32/0.12 = 8/3$, positive when $s > 8/3$, and zero when $s = 8/3$, making $g(8/3)^5$ the maximum value. Plugging in $s = 8/3$ gives a bound of $16/27 \leq \boxed{0.5926}$.

SAMPLE AVERAGES

The advantage of Chernoff's bound for sums extends to sample averages. For $S = S_1 + \cdots + S_n$ where the S_i are iid, the sample average S/n has

$$\mathbb{P}(S/n \geq a) = \mathbb{P}(S \geq an)$$
$$\leq \frac{\text{mgf}_{S_1}(t)^n}{\exp(tan)}$$
$$= \left[\frac{\text{mgf}_{S_1}(t)}{\exp(ta)}\right]^n.$$

In other words, if S_1 has a Chernoff bound γ on its tail, then the tail of the sample average of n iid copies of S_1 is γ^n.

ENCOUNTERS

287. CHERNOFF FOR A POISSON

For $X \sim \text{Pois}(21.3)$, use Chernoff's inequality to bound $\mathbb{P}(X \geq 30)$.

288. MORE CHERNOFF FOR POISSONS

For $N \sim \text{Pois}(10)$, use Chernoff's inequality to bound $\mathbb{P}(X \leq 4)$.

289. CHERNOFF FOR GAMMAS

Using $t = 0.47$ in Chernoff's bound, give an upper bound for $R \sim \text{Gamma}(13, 1.4)$ of $\mathbb{P}(R \geq 14)$.

290. MORE CHERNOFF FOR GAMMAS

Using $t = 0.8$ in Chernoff's bound, give an upper bound for $T_{15} \sim \text{Gamma}(15, 2.3)$ of $\mathbb{P}(T_{15} \geq 10)$.

291. CHERNOFF GIVEN THE MGF

Suppose that X has $\text{mgf}_X(0.2)/\exp(4(0.2)) \leq 0.6$, and $X_1, \ldots, X_{10}$ are iid X. What is a bound on
$$\mathbb{P}\left(\frac{X_1 + \cdots + X_{10}}{10} \geq 4\right)?$$

292. MORE CHERNOFF GIVEN THE MGF

Suppose that Y has
$$\text{mgf}_Y(1.3)/\exp(6.1(1.3)) \leq 0.45,$$
and $Y_1, \ldots, Y_8$ are iid Y. What is a bound on
$$\mathbb{P}\left(\frac{Y_1 + \cdots + Y_8}{8} \geq 6.1\right)?$$

CHAPTER 26: CHERNOFF INEQUALITIES

293. CHERNOFF FOR UNIFORMS

For $U_1, \ldots, U_n$ iid Unif($[0,1]$), consider

$$P((U_1 + \cdots + U_{10}) \geq 6).$$

a. Use Wolfram Alpha to find the value of t that minimizes the Chernoff bound for this probability.

b. Find the best Chernoff bound for this probability.

294. CHERNOFF FOR DISCRETE UNIFORMS

Let $X \sim$ Unif($\{2, 3, 5\}$). For $X_1, \ldots, X_{20}$ iid X, consider $\mathbb{P}(X_1 + \cdots + X_n \geq 75)$.

a. Use Wolfram Alpha to find the value of t that minimizes the Chernoff bound for this probability.

b. Find the best Chernoff bound for this probability.

Chapter 27: The Hypergeometric Distribution

In the bag of loot there were five gold pieces and seven silver pieces. The adventurers decided to split the twelve pieces among their three person party by allowing each member to take four pieces from the bag. What is the chance that the first member to draw took two gold and two silver?

What is the chance that the last person to draw was left with two gold and two silver?

The Hypergeometric Distribution

Recall that if $B_1, B_2, \ldots$ are iid $\mathrm{Bern}(p)$, then call this a *Bernoulli process*, and $\{i : B_i = 1\}$ is a Bernoulli point process with parameter p.

This gives rise to the binomial distribution:

$$B_1 + \cdots + B_n \sim \mathrm{Bin}(n, p),$$

the geometric distribution

$$\inf\{i : B_1 + \cdots + B_i = 1\} \sim \mathrm{Geo}(p),$$

and the negative binomial distribution

$$\inf\{i : B_1 + \cdots + B_i = r\} \sim \mathrm{NegBin}(r, p),$$

> **Numbers of Bernoulli points**
> Because it arises so frequently, throughout this chapter let
> $$N_k = B_1 + \cdots + B_k.$$

It also gives rise to one more distribution not discussed earlier. This distribution looks at the number of points in the first k positions given the number of points in the first n positions.

> **D94** For a Bernoulli process (of any $p \in (0,1)$) with $n \in \{1, 2, \ldots\}, k \in \{1, \ldots, n\}$ and $\ell \in \{1, \ldots, n\}$, consider the number of points in the first k positions conditioned on the number of points in the first n positions. This has a *hypergeometric distribution*. For
> $$N_r = B_1 + \cdots + B_r,$$
> write
> $$[N_k \mid N_n = \ell] \sim \mathrm{HyperGeo}(n, k, \ell).$$

For $H \sim \mathrm{HyperGeo}(n, k, \ell)$, the density of H can be found by using the conditional probability formula.

> **F102** For $H \sim \mathrm{HyperGeo}(n, k, \ell)$,
> $$\mathbb{P}(H = i) = \frac{\binom{k}{i}\binom{n-k}{\ell-i}}{\binom{n}{\ell}}.$$

Proof. Let $i \in \{1, \ldots, \min(k, \ell)\}$. Then

$$\mathbb{P}(H = i) = \mathbb{P}(N_k = i \mid N_n = \ell)$$
$$= \frac{\mathbb{P}(N_k = i, N_n = \ell)}{\mathbb{P}(N_n = \ell)}$$
$$= \frac{\mathbb{P}(N_k = i, N_n - N_k = \ell - i)}{\mathbb{P}(N_n = \ell)}$$
$$= \frac{\mathbb{P}(N_k = i)\mathbb{P}(N_n - N_k = \ell - i)}{\mathbb{P}(N_n = \ell)}$$
$$= \frac{\binom{k}{i}\binom{n-k}{\ell-i}}{\binom{n}{\ell}},$$

where $p^i(1-p)^{n-i}$ appeared in both the numerator and the denominator, and so was canceled. $\square$

This calculation did not rely on $B_1, \ldots, B_k$ being the points considered, any set $\{a_1, \ldots, a_k\}$ that containing k positions in $\{1, \ldots, n\}$ would work. This gives a generalization of the last fact.

> **F103** Let $S = \{1, \ldots, n\}$ and $A \subseteq S$ with $\#(S) = n$ and $\#(A) = k$. Then for any $\ell \in \{1, \ldots, n\}$,
> $$\left[\sum_{a \in A} B_a \,\middle|\, \sum_{i \in S} B_i = \ell\right] \sim \mathsf{HyperGeo}(n, k, \ell)$$

The proof is similar to the last fact.

SOLVING THE STORY

Consider the bag of twelve coins. Visualize the inside of the bag as a Bernoulli process, where a 1 means the coin is gold, and a 0 means that the coin is silver.

The fact that out of the twelve coins, 5 of the coins are gold and 7 are silver, means that
$$B_1 + \cdots + B_{12} = 5.$$

So if the first four coins are drawn out of the bag, then
$$[N_4 \mid N_{12} = 10] \sim \mathsf{HyperGeo}(12, 4, 10).$$

That means
$$\mathbb{P}(N_4 = 2 \mid N_{12} = 10) = \frac{\binom{4}{2}\binom{7}{2}}{\binom{12}{4}},$$

which simplifies to 14/33 and is about $\boxed{0.4242}$.

The story of the bag also asks about the probability that for the last four coins drawn from the bag, what is the chance that two gold are selected. This is still
$$\mathbb{P}(B_9 + \cdots + B_{12} \mid B_1 + \cdots + B_{12} = 5),$$

which has the same distribution as before. Hence it is still $\boxed{0.4242}$. Remember, whether looking at the first four coins or the last four coins picked, the distribution will be exactly the same!

UNIFORM POINTS

A hypergeometric random variable gives us the distribution of the sum of the points over k positions when the points over n positions is known to be ℓ. But what about the locations of those points?

For $B = \{i : B_i = 1\}$, and $S = \{1, \ldots, n\}$ this is the set $B \cap S$. Again, condition on $\#(B) = \ell$. Then what is the distribution of B?

Well, each Bernoulli looks exactly like another, so intuitively each subset of S of size ℓ should be equally likely to equal B. This intuition is correct!

> **F104** Let $S = \{1, \ldots, n\}$, $B_1, B_2, \ldots$ iid $\mathsf{Bern}(p)$ for $p \in (0, 1)$, and $B = \{i : B_i = 1\}$. Then
> $$[B \mid \#(B) = \ell] \sim \mathsf{Unif}(\{b \subseteq S : \#(b) = \ell\})$$

Proof. Let b be a subset of S with ℓ elements. Then
$$\mathbb{P}(B = b \mid \#(B) = \ell) = \frac{\mathbb{P}(B = b, \#(B) = \ell)}{\mathbb{P}(\#(B) = \ell)}$$
$$= \frac{\mathbb{P}(B = b)}{\mathbb{P}(\#(B) = \ell)}$$
$$= \frac{p^\ell(1-p)^{n-\ell}}{\binom{n}{\ell}p^\ell(1-p^{n-\ell})}$$
$$= 1/\binom{n}{\ell}.$$

There are $\binom{n}{\ell}$ such sets b, so the distribution must be uniform over these sets. □

Note that p cancels out as with the hypergeometric, so this fact holds regardless of the value of p! Once the number of points is fixed, the position of those points is uniform over possible positions, whether originally each Bernoulli was likely or unlikely to occur.

Also, conditioned on $B_1 + \cdots + B_n$, each B_i is still a Bernoulli random variable with the same mean.

> **F105** For $n \in \{1, 2, \ldots\}$, $\ell \in \{0, \ldots, n\}$, and $i \in \{1, \ldots, n\}$,
> $$[B_i \mid B_1 + \cdots + B_n = \ell] \sim \operatorname{Bern}\left(\frac{\ell}{n}\right).$$

Proof. This follows from the hypergeometric distribution with $k = 1$:

$$\mathbb{P}(B_i = 1 \mid N_n = \ell) = \frac{\binom{1}{1}\binom{n-1}{\ell-1}}{\binom{n}{\ell}}$$
$$= \frac{(n-1)!}{(n-\ell)!(\ell-1)!} \cdot \frac{\ell!(n-\ell)!}{n!}$$
$$= \frac{\ell}{n}.$$

□

Of course, even if they are identically distribution conditioned on $B_1 + \cdots + B_n = \ell$, they are not independent. To see that, consider the covariance between B_i and B_j for $i \neq j$.

> **F106** For $n \in \{1, 2, \ldots\}$, $\ell \in \{0, \ldots, n\}$, and $i \in \{1, \ldots, n\}$, for $i \neq j$, let s_ℓ be the event that $B_1 + \cdots + B_n = \ell$. Then
> $$\operatorname{cov}([B_i \mid s_\ell], [B_j \mid s_\ell]) = \frac{\ell(\ell-1)}{n(n-1)} - \frac{\ell^2}{n^2}.$$

Proof. Because B_i and B_j are 0 or 1, so is $B_i B_j$, so this is also a Bernoulli random variable! Moreover,

$$\mathbb{P}(B_i B_j = 1 \mid s_\ell) = \mathbb{P}(B_i + B_j = 2 \mid s_\ell)$$
$$= \frac{\binom{2}{2}\binom{n-2}{\ell-2}}{\binom{n}{\ell}}$$
$$= \frac{(n-2)!}{(\ell-2)!(n-\ell)!} \cdot \frac{\ell!(n-\ell)!}{n!}$$
$$= \frac{\ell(\ell-1)}{n(n-1)}$$

So
$$c = \operatorname{cov}([B_i \mid s_\ell], [B_j \mid s_\ell])$$
$$= \mathbb{E}[B_i B_j \mid s_\ell] - \mathbb{E}[B_i \mid s_\ell]\mathbb{E}[B_j \mid s_\ell]$$
$$= \frac{\ell(\ell-1)}{n(n-1)} - \frac{\ell^2}{n^2}.$$

□

Hence the covariance is slightly negative, which is expected given that when $B_i = 1$, that slightly lowers the chance that $B_j = 1$ given that the sum of the Bernoullis is fixed.

MOMENTS OF HYPERGEOMETRICS

First consider the first moment of the hypergeometric, that is, the mean.

> **F107** For $H \sim \operatorname{HyperGeo}(n, k, \ell)$,
> $$\mathbb{E}[H] = \frac{k\ell}{n}.$$

Proof.

$$\mathbb{E}[H] = \mathbb{E}[B_1 + \cdots + B_k \mid N_n = \ell]$$
$$= \mathbb{E}[B_1 \mid N_n = \ell] + \cdots + \mathbb{E}[B_k \mid N_n = \ell]$$
$$= k\mathbb{E}[B_1 \mid N_n = \ell]$$
$$= k\ell/n$$

as desired. □

Using our variance formula for sums gives the following.

> **F108** For $H \sim \operatorname{HyperGeo}(n, k, \ell)$,
> $$\operatorname{var}(H) = \frac{k\ell}{n}\left[\frac{(n-k)(n-\ell)}{n(n-1)}\right].$$

Proof. Start with

$$\operatorname{var}(H) = \operatorname{var}(N_k \mid N_n = \ell)$$
$$= \operatorname{var}(B_1 + \cdots + B_k \mid N_n = \ell).$$

The variance of sums formula says that $\mathrm{var}(H)$ is

$$\mathrm{var}(H) = s_1 + s_2,$$

where

$$s_1 = \sum_{i=1}^{k} \mathrm{var}(B_i \mid N_n = \ell)$$

and

$$s_2 = \sum_{i \neq j} \mathrm{cov}([B_i \mid N_n = \ell], [B_j \mid N_n = \ell]).$$

Consider s_1, the sum of variances first. Each B_i conditional on $N_n = \ell$ is a Bernoulli random variable with mean ℓ/n. Hence the variance of one is

$$\frac{\ell}{n}\left(1 - \frac{\ell}{n}\right),$$

and the variance of the sum of k identically distributed random variables is

$$s_1 = k\frac{\ell}{n}\left(1 - \frac{\ell}{n}\right).$$

Similarly,

$$\mathrm{cov}([B_i \mid s_\ell], [B_j \mid s_\ell]) = \frac{\ell(\ell-1)}{n(n-1)} - \frac{\ell^2}{n^2}$$
$$= \frac{\ell}{n}\left[\frac{\ell-1}{n-1} - \frac{\ell}{n}\right].$$

There are $k(k-1)$ terms in the covariance sum, so

$$s_2 = \frac{k\ell}{n}\left[\frac{(k-1)(\ell-1)}{n-1} - \frac{(k-1)\ell}{n}\right].$$

This means

$$s_1 + s_2 = \frac{k\ell}{n}\left[1 - \frac{\ell}{n} + \frac{(k-1)(\ell-1)}{n-1} - \frac{(k-1)\ell}{n}\right]$$
$$= \frac{k\ell}{n}\left[1 + \frac{(k-1)(\ell-1)}{n-1} - \frac{k\ell}{n}\right]$$
$$= \frac{k\ell}{n} \cdot \frac{(n-k)(n-\ell)}{n(n-1)},$$

as desired. □

ANOTHER VIEW

The mean and variance formulas are the same if k and ℓ are swapped. It turns out this symmetry goes deeper:

$$\mathrm{HyperGeo}(n, k, \ell) = \mathrm{HyperGeo}(n, \ell, k).$$

To understand why this is true, suppose the objects to be drawn *and* the objects that are drawn are both viewed as being chosen uniformly at random. This gives a viewpoint of the hypergeometric with a nice symmetry property.

F109 Let $S = \{1, \ldots, n\}$. Suppose A is a randomly chosen subset of S of size k, and B is a randomly chosen subset of S of size ℓ. Then

$$\#(A \cap B) \sim \mathrm{HyperGeo}(n, k, \ell).$$

Proof. Let $i \in \{1, \ldots, k\}$, then

$$\mathbb{P}(\#(A \cap B) = i) = \mathbb{E}[\mathbb{I}(\#(A \cap B) = i)]$$
$$= \mathbb{E}[\mathbb{E}[\mathbb{I}(\#(A \cap B)) = i \mid A]]$$
$$= \mathbb{E}[f_H(i)]$$
$$= f_H(i).$$

□

In other words, whatever is chosen for the set A it holds that

$$\left[\sum_{a \in A} B_a \;\Big|\; \sum_{s \in S} B_s = \ell\right]$$

has a hypergeometric distribution. Choosing the A uniformly does not change that.

One of the the nice things about this view of a hypergeometric is that the symmetry between k and ℓ is immediately apparent.

F110

$$\mathrm{HyperGeo}(n, k, \ell) \sim \mathrm{HyperGeo}(n, \ell, k)$$

Proof. This follows from
$\#(A \cap B) = \#(B \cap A)$. □

It also gives us a density formulation for the hypergeometric that is explicitly symmetric in k and ℓ.

> **F111** For $H \sim \text{HyperGeo}(n, k, \ell)$, and $i \in \{0, \ldots, \min(k, \ell)\}$,
> $$\mathbb{P}(H = i) = \binom{n}{i} \frac{\binom{n-i}{k-i, \ell-i, n-k-\ell+i}}{\binom{n}{k}\binom{n}{\ell}}.$$

Proof. The denominator $\binom{n}{i}\binom{n}{k}$ is the number of ways to choose the subsets A and B. Then $\binom{n}{i}$ is the number of ways to choose AB where $\#(AB) = i$, and then
$$\binom{n-i}{k-i, \ell-i, n-k-\ell+i}$$
is the number of ways to choose the $k-i$ elements of $A \setminus AB$ and the $\ell-i$ elements of $B \setminus AB$ from among $\{1, \ldots, n\} \setminus AB$. □

Since $\binom{n}{a,b,c} = \binom{n}{b,a,c}$, this expression for $\mathbb{P}(H = i)$ is the same if k and ℓ are swapped.

ENCOUNTERS

295. Those crazy eights

In a standard 52 deck of cards, there are four cards with rank 8 (the 8 of hearts, the 8 of spades, the 8 of diamonds, and the 8 of clubs).

If seven cards are dealt out uniformly at random from the deck, what is the chance that exactly two are rank 8?

296. Pick a card, any card

In a standard 52 deck of cards, there are 13 hearts.

a. If five cards are chosen uniformly at random, what is the chance that all the cards are hearts?

b. If five cards are chosen uniformly at random, what is the chance that exactly three of the cards are hearts?

297. A bag of marbles

A bag contains five red and ten blue marbles. If four marbles are selected at random, what is the chance that exactly three are red?

298. Strains

There are 23 strains of a particular virus active in the world today. Three of them require immediate treatment. If ten patients have the virus, and are equally likely to have any strain, what is the chance that exactly one patient has one of the strains that require immediate treatment?

299. A box of screws

A box of screws contains 40 type A and 60 type B screws. If a group of 30 screws is chosen uniformly at random from the box, then what is the chance that the last screw chosen is type A?

300. A bag of tokens

A bag contains seven blue tokens and six green tokens. If 6 tokens are taken out of the bag without replacement uniformly at random, what is the chance that there are an equal number of blue and green tokens in the sample?

301. The mission

Three out of eight members of the Ranger's Guild are Wood Elves. If five of the members are chosen for a secret mission uniformly at random, what is the chance that exactly two are Wood Elves?

302. BACK TO THE ASSEMBLY LINE

Suppose that an assembly line creates 20 items in an hour. Unknown to the tester, 4 of these 20 items are defective. In that case, if the tester tests 5 different items chosen uniformly from the 20, what is the chance that the third item tested is defective?

303. THE FOREST

There are believed to be 20 deer in a forest. During one survey, 5 of the deer are tagged. During the second survey, 11 of the deer are tagged. If the tagging of the deer is random, let T be the number of deer tagged twice.
a. What is $\mathbb{E}[T]$?
b. What is $\text{sd}(T)$?

304. TURTLES ALL THE WAY DOWN

From a group of 50 turtles being studied by ecologists, 7 are chosen uniformly at random to be tagged. A second survey of the same turtles tags 10 of them uniformly at random. Let T be the number of turtles tagged twice.
a. What is $\mathbb{E}[T]$?
b. What is $\text{sd}(T)$?

Chapter 28: Poisson point processes over general spaces

GIANT RATS HAVE INFESTED THE cellar! The cellar is 120 square feet in size. It has a left side with 70 square feet, and a right side with 50 square feet. If the locations of the rats follow a Poisson point process of rate 0.1 per square foot, what is the chance that a given rat is located on the left side of the cellar?

D95 A *Poisson point process* over a set $S \in \mathbb{R}^n$ with Lebesgue measure and constant rate λ is defined as follows.

a. Let $N \sim \text{Pois}(\lambda \operatorname{Leb}(S))$.

b. Let $P = \{X_1, \ldots, X_N\}$ be uniformly distributed over the space S.

Write $P \sim \text{PPP}(S, \lambda)$.

In the definition, uniformity of the points is part of the construction. In the cellar, the Lebesgue measure of the region is 120 square feet, and so the number of giant rats is Poisson distributed with a single parameter $(120)(0.1) = 12$. That parameter will be the average number of rats in the cellar, but there might be more or less.

The following is a picture of the cellar in the story.

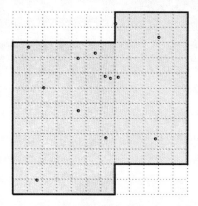

The uniform point of view

In the last chapter, it was observed that in a Bernoulli point process, the number of points in a given set S was binomial with parameters $\#(S)$ and parameter p. Conditioned on the number of points, they were distributed uniformly on S without replacement.

This gives a new way of defining a Poisson point process on state spaces more general than just $[0, \infty)$. This is a two step process. First, decide the number of points using a Poisson point process. Second, place the points down uniformly.

In the picture above, there were 13 rats in the cellar. Of these, 9 ended up on the left side, and 4 ended up on the right side. To answer the question in the story, given that each rat was dropped uniformly in the cellar, there is a $70/120 \approx 58.33\%$ chance of landing on the left side. This is because the probability that $X \sim \text{Unif}(S)$ falls in A is the measure of A divided by the measure of S.

> **E69** Suppose that $X \sim \text{PPP}(S, 2.3)$, where S is a three dimensional space of volume 84. Then $A \subseteq S$ has volume 10. What is the chance that any particular point in X falls into A?
> **Answer.** Any particular point is uniformly distributed, so this is just
> $$\frac{10}{84} \approx \boxed{0.1190}.$$

From uniform to binomial

Consider the distribution of the number of rats on the left hand side conditioned on having n points in the process. Because each of the rat locations is uniform, this will have a binomial distribution. The probability of falling in the left hand side is proportional to the area of the left hand side.

> **F112** For $P \sim \text{PPP}(S, \lambda)$,
> $$[\#(P \cap A) \mid \#(P \cap S) = n] \sim \text{Bin}(n, \mu(AS)/\mu(S)).$$

> **E70** Let $S = [0,1] \times [0,2]$ and $A = [0.5, 1] \times [0.5, 1.5]$. Suppose P is a Poisson point process over S with 5 points. what is the distribution of $\#(P \cap A)$?
> **Answer.** The measure (area) of $A \subseteq S$ is $(1 - 0.5)(1.5 - 0.5) = 0.5$. The measure of S is $(1 - 0)(2 - 0) = 2$. Hence the distribution of $\#(P \cap A)$ is
> $$\boxed{\text{Bin}(5, 0.2500)}.$$

So for Bernoulli processes, the number of points that fall into a particular subset has a hypergeometric distribution, but for Poisson processes, the number of points that fall into a particular subset will be binomial! No need to create a new distribution.

Encounters

305. Back to the cellar

In the cellar from the story the space is a polygon with vertices $(0,0), (0,12), (12,12), (12,2), (7,2)$ and $(7,0)$.
a. What is the chance that there are at least 12 rats in the cellar?
b. Given a point that marks a rat's location, what is the chance that the point has second coordinate at least 2?
c. What is the chance that there are no points with second coordinate less than 2?

306. The disease

An epidemiologist starts with a basic model where the infected in a town follow a Poisson point process. The part of town south of the river is of size 5 square miles, while the part north of the river is of size 6 square miles.
a. If the model is correct, what is the chance that a given infection location is south of the river?
b. If the model is correct, and the rate of infection is about 10 per square mile, what is the average number of infections north of the river?

307. The restaurant

Customers arrive to a restaurant at times modeled by a Poisson point process of rate 15 per hour. If 10 customers arrive in the first hour, what is the chance that exactly 5 customers arrive in the first half-hour?

308. The book

Sadly most books contain typos. Suppose a book has 30 chapters of equal length. If typos in a book are modeled as a Poisson point process of rate 10 per chapter, and there are 278 typos total in the book, what is the chance that there are no typos in the first chapter at all?

309. AN ABSTRACT SPACE

Suppose S is a space with measure $\mu(S) = 15.2$. A set $A \subseteq S$ has $\mu(A) = 11.4$. Given a point $x \in S$ in the Poisson point process of constant rate $\lambda = 2.1$ over S, what is the chance that $x \in A$?

310. MORE ABSTRACT FUN!

Suppose that P is a Poisson point process of constant rate λ over a space W that has measure $\mu(W) = 200$. Given a subset $A \subseteq W$ with $\mu(A) = 50$, find the chance that if P has 10 points in the space, exactly five of the points land in A.

Chapter 29: Transforming Multivariate Random Variables

COMPLEX ARE THE MINDS OF Wizards, and the model this Wizard had built was certainly complicated. The Wizard started with two random variables, U and V iid standard normals. Then these were used to build two new random variables,

$$W = U^2$$
$$Y = U - V.$$

The Wizard knew that W and Y were not independent, after all, $\mathbb{P}(Y \leq \sqrt{W}) = 1$. What, the Wizard wondered, would be the joint density of the pair (W, Y)?

Multivariate Transformations

While the Wizard's thoughts might be entirely theoretical, multivariate transformations are relatively common in actual statistical analyses. So it is necessary to develop a method for transforming the joint density of the original random variables into the joint density of the new variables.

To begin, consider what happens in one dimension.

Nonlinear Transformations in One Dimension

Start with a linear transformation: for X with density f_X with respect to Lebesgue measure, the density of $aX + b$ (where $a \neq 0$) is

$$f_{aX+b}(s) = \frac{1}{a} f_X\left(\frac{s-b}{a}\right).$$

Now consider an example of a *nonlinear* transformation. Suppose that $U \sim \text{Unif}([0,1])$ and $W = U^2$. Since $U \geq 0$, this transformation is 1-1: given $W \in [0,1]$ there is exactly one value of U that maps to W. In this situation, one can find the density of W using the linear approximation of the mapping. That is, use the derivative to calculate the density.

Before showing this, it will help to know a bit more about densities. A density f of random variable X has the property that for all $a < b$,

$$\mathbb{P}(X \in [a,b]) = \int_a^b f(x) \, d\mu(x).$$

It turns out that it is not necessary to show this for every $a < b$. Instead, it is enough that it is true for at least one interval containing x for every x.

F113 Suppose that for every x, there exists an interval $[a, b]$ such that $a < x < b$ and

$$\mathbb{P}(X \in [a,b]) = \int_a^b f(x) \, dx.$$

Then f is a density of X.

The proof is usually shown in a real analysis course.

F114 Suppose that $y = h(x)$ is a 1-1 mapping with nonzero, continuous derivative that is either always positive or always negative over the domain of h. Then for X with density f_X with respect to Lebesgue measure with probability 1 of being in the domain of h, the random variable $Y = h(X)$ has density

$$f_Y(y) = \frac{1}{|h'(x)|} f_X(x),$$

where $x = h^{-1}(y)$.

Proof. Because h has always positive or always negative derivative over $[a, b]$, it is either strictly increasing or strictly decreasing over $[a, b]$. Let $[c, d]$ be any subset of $[a, b]$. Then the goal is to find $p = \mathbb{P}(Y \in [c, d])$.

$$\begin{aligned} p &= \mathbb{P}(Y \in [c, d]) \\ &= \mathbb{E}[\mathbb{I}(h(X) \in [c, d])] \\ &= \int \mathbb{I}(h(x) \in [c, d]) f_X(x) \, dx \\ &= \int_{x:h(x)\in[c,d])} f_X(x) \, dx \end{aligned}$$

Now make the substitution $y = h(x)$ so $dy/dx = h'(x)$, and $dx = (1/h'(x)) \, dy$. Suppose $h'(x) > 0$ for all $x \in [a, b]$. Then

$$p = \int_{y=h^{-1}(c)}^{h^{-1}(d)} f_X(h^{-1}(x))/h'(x) \, dx$$

Since this holds for a nontrivial interval around x, $f_X(h^{-1}(x))/h'(x) \, dx$ must be the density of $h(X)$.

Similarly, if $h'(x) < 0$ for all points in the interval $[c, d]$, then $h(x)$ is decreasing, and

$$\begin{aligned} p &= \int_{y=h^{-1}(d)}^{h^{-1}(c)} f_X(h^{-1}(y))/h'(x) \, dx \\ &= \int_{y=h^{-1}(c)}^{h^{-1}(d)} -(1/h'(x)) f_X(h^{-1}(x)) \, dx. \end{aligned}$$

Note when $h'(x) < 0$, $-h'(x) = |h'(x)|$. Therefore, in either case,

$$p = \int_{y=h^{-1}(c)}^{h^{-1}(d)} (1/|h'(x)|) f_X(h^{-1}(y)) \, dx.$$

This means that $(1/|h'(h^{-1}(y))|) f_X(h^{-1}(y))$ is the density of f_Y. □

E71 For $U \sim \text{Unif}([0, 1])$, let $W = \ln(1/U)$. Find the density of W.

Answer. Here $h(u) = \ln(1/u)$ is a strictly decreasing function of u over the domain with continuous derivative inside the domain. For $w = \ln(1/u)$, solving gives $u = \exp(-w)$. Also, $h'(u) = -u^{-2} \cdot (1/u)^{-1} = -u^{-1}$. Hence

$$\begin{aligned} f_W(w) &= (1/|h'(\exp(-w))|) f_U(\exp(-w)) \\ &= |1/(-1/\exp(-w))| \mathbb{I}(\exp(-w) \in [0, 1]) \\ &= \exp(-w) \mathbb{I}(w \in [0, \infty)) \end{aligned}$$

which is the density of a standard exponential as expected.

HIGHER DIMENSIONAL TRANSFORMS

When the transformation is multivariate, the idea is the same, although the execution is a bit more complicated.

Suppose that F is a function that takes n real numbers as inputs $x = (x_1, x_2, \ldots, x_n)$ and returns n real numbers as outputs $y = (y_1, \ldots, y_n)$. Write $F : \mathbb{R}^n \to \mathbb{R}^n$, and $y = F(x)$.

Because F has n outputs, one way to view F is as a compilation of n different functions:

$$F(x) = (F_1(x), F_2(x), \ldots, F_n(x)).$$

In the univariate case, the derivative was used to create a linear approximation to the transformation. In the multivariate case, the *Jacobian matrix* is used.

F115 For $x \in \mathbb{R}^n$ and multivariate function $F(x) = (F_1(x), F_2(x), \ldots, F_n(x))$, the Jacobin matrix of F is an n by n matrix defined by

$$J_F = \begin{pmatrix} \frac{\partial F_1}{\partial y_1} & \cdots & \frac{\partial F_n}{\partial y_n} \\ \vdots & \vdots & \vdots \\ \frac{\partial F_n}{\partial y_1} & \cdots & \frac{\partial F_n}{\partial y_n} \end{pmatrix}$$

CHAPTER 29: TRANSFORMING MULTIVARIATE RANDOM VARIABLES

Then for a small $h \in \mathbb{R}^n$,
$$F(x+h) \approx F(x) + J_F(x)h,$$
making it a good linear approximation.

In one dimension, the factor of the volume change is just the absolute value of the derivative. In higher dimensions, things are more complex. A function of an n by n matrix called the *determinant* is used to calculate the volume change.

This leads to the following result.

> **F116** Let $F : \mathbb{R}^n \to \mathbb{R}^n$ be an invertible transformation with Jacobian J_F. Suppose $X = (X_1, \ldots, X_n)$ has density f_X and let $Y = F(X)$. Then the density of Y is
> $$f_Y(y) = \frac{1}{|\det J_F(x)|} f_X(x),$$
> where $x = F^{-1}(y)$.

JACOBIAN ALERT!
The determinant of the Jacobian matrix is also sometimes called the Jacobian. Mathematicians have never quite settled on whether the term Jacobian should be the matrix or the determinant of the matrix.

SOLVING THE STORY

In the story, U and V are iid Unif($[0,1]$). So
$$f_{(U,V)}(u,v) = \mathbb{I}((u,v) \in [0,1]^2).$$

The transformation is
$$(W,Y) = F(U,V),$$
where
$$(w,y) = F(u,v) = (u^2, u-v).$$

Some quick differentiating gives the Jacobian:
$$J = \begin{pmatrix} 2u & 0 \\ 1 & -1 \end{pmatrix}$$

which makes the absolute value of the derivative:
$$|-2u - 1 \cdot 0| = 2u$$
since this will only be applied when $u \in [0,1]$.

Hence
$$f_{(W,Y)}(w,y) = \frac{1}{2u} \mathbb{I}((u,v) \in [0,1]^2).$$

Of course, the goal is to put the right hand side in terms of w and y. If $u \in [0,1]$, then $w \in [0,1]$. Given $u = \sqrt{w}$ and v in $[0,1]$ makes the difference $y = u - v \in [\sqrt{w}, \sqrt{w} - 1]$.

Finally, since $w = u^2$, $u = \sqrt{w}$. Hence the final answer is
$$\boxed{f_{(W,Y)}(w,y) = \frac{1}{2\sqrt{w}} \mathbb{I}(0 \leq w \leq 1, \sqrt{w} - 1 \leq y \leq \sqrt{w})}$$

POLAR COORDINATES

One of the most common $F : \mathbb{R}^2 \to \mathbb{R}^2$ is the transformation from rectangular coordinates to polar coordinates, which works as follows.

> **D96** For $(x,y) \in \mathbb{R}^2$, say that (r,θ) represents the point in **polar coordinates** if $r \geq 0$, $\theta \in [0, \tau)$, and
> $$x = r\cos(\theta), \quad y = r\sin(\theta).$$

CIRCULAR REASONING
This transformation is only invertible if θ is restricted to lie in $[0, \tau)$. Otherwise (r, θ) and $(r, \theta + \tau)$ represent the same point.

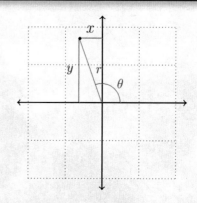

CHAPTER 29: TRANSFORMING MULTIVARIATE RANDOM VARIABLES

The Jacobian is a bit tricky to calculate here given the inverse sin and cos change as the point crosses over quadrants, so here the resulting transformation is just given in differential form.

> **F117** For the transformation from rectangular to polar coordinates,
> $$dx\, dy = r \cdot \mathbb{I}(r \geq 0)\mathbb{I}(\theta \in [0, \tau))\, dr\, d\theta.$$

E72 Let (Z_1, Z_2) be two iid standard normals. What is the density of (R, Θ) in polar coordinates?

Answer. Note that

$$\mathbb{P}(Z_1 \in dx, Z_2 \in dy)$$
$$= (f_{Z_1}(x)\, dx)(f_{Z_2}(y)\, dy)$$
$$= \frac{1}{\tau} \exp\left(-\frac{x^2 + y^2}{2}\right) dx\, dy$$
$$= \frac{1}{\tau} \exp\left(-\frac{r^2}{2}\right) r \mathbb{I}(r \geq 0)\mathbb{I}(\theta \in [0, \tau))\, dr\, d\theta$$
$$= \left[\frac{1}{\tau}\mathbb{I}(\theta \in [0, \tau))\, d\theta\right] \left[r \exp(-r^2/2)\, dr\right]$$

That means the joint density factors into a density for Θ that is uniform over $[0, \tau)$, and a density for R that is the standard Rayleigh distribution, so R and Θ are independent when they are the polar coordinate form of two independent standard normal random variables!

ENCOUNTERS

311. ONE DIMENSIONAL TRANSFORM

Suppose X has pdf (for $x > 0$)

$$f_X(x) = \frac{1}{x\sqrt{\tau}} \exp(-\log(x)^2/2).$$

Let $Y = \log(X)$.
a. What is the density of Y?
b. What is the distribution of Y?

312. TRANSFORMING AN EXPONENTIAL

Let $T \sim \text{Exp}(1)$ be a standard exponential. What is the density of $R = \sqrt{2T}$?

313. TWO DIMENSIONAL TRANSFORM

Suppose $f(x, y) = x^2 y$. If $X \sim \text{Unif}([0, 1])$ and $Y \sim \text{Unif}([0, 1])$, what is the density of $S = f(x^2, y)$?

314. ANOTHER 2D TRANSFORM

If U and V are independent standard uniforms, find the density of

$$X = \frac{U - V}{U + V}.$$

Chapter 30: Encounters Resolved

EREIN LIE TALES OF PROBLEMS vanquished! Not every encounter in the text has been resolved, but perhaps the seeker of knowledge will find some light on their path by examining the stories told in this chapter.

What is Probability?

1. Basic Indicators

What is $\mathbb{I}(4^2 > 10)$?

Solution
Since $4^2 = 16 > 10$ is true, the answer is $\boxed{1}$.

3. Basic Probabilities

What is $\mathbb{P}(4^2 > 10)$?

Solution
Since $4^2 = 16 > 10$ is true, the answer is $\boxed{1}$.

5. Indicator Functions

Suppose $f(x) = \mathbb{I}(|x| > 4)$.
a. What is $f(2)$?
b. What is $f(-2)$?
c. What is $f(5)$?
d. What is $f(-5)$?

Solution
a. Since $(|2| > 4) = \mathsf{F}$, this is $\boxed{0}$.
b. Also, $(|-2| > 4) = \mathsf{F}$, this is again $\boxed{0}$.
c. Here $(|5| > 4) = \mathsf{T}$, so this is $\boxed{1}$.
d. This is $(|-5| > 4) = \mathsf{T}$, so again is $\boxed{1}$.

7. Graphing Indicator Functions

Indicator functions, when used to make functions, can be graphed.
a. Suppose $f(x) = \mathbb{I}(|x| > 4)$. Graph $f(x)$.
b. Suppose $g(x) = (|x|/4)\mathbb{I}(|x| > 4)$. Graph $g(x)$.

Solution
a. This function is constant 1 whenever $x < -4$ or $x > 4$, and 0 whenever $-4 \leq x \leq 4$. So it looks like:

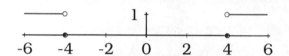

b. When the function is nonzero, the line being graphed is $|x|/4$. So that looks like this:

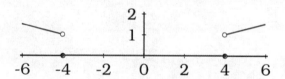

9. Disjoint Reals

Suppose X is a real number. State if the following events are disjoint or not.
a. $(X \leq 3)$ and $(X \geq 4)$.
b. $(X \leq 5)$ and $(X \geq 3)$.
c. $(X \leq 3)$ and F.

Solution
a. There is no real number that is both at most 3 and at least 4, so these intervals are $\boxed{\text{disjoint}}$.
b. The number $X = 4$ is both at most 5 and at least 3, so these are $\boxed{\text{not disjoint}}$.
c. Since F is always F, at most one of the statements is true, making them $\boxed{\text{disjoint}}$.

11. Truth or dare?

What is $\mathbb{P}(10 < 20)$?

Solution

This is a true statement, so the probability is $\boxed{1}$.

13. Counting true statements

Suppose s_1, s_2, and s_3 are disjoint statements. What is the largest that $\mathbb{I}(s_1) + \mathbb{I}(s_2) + \mathbb{I}(s_3)$ can be?

Solution

The sum of indicators of disjoint events can be at most $\boxed{1}$.

15. Two dice

If A is the roll of a fair six-sided die, (so $A \sim d6$) and B is the roll of a fair four-sided die (so $B \sim d4$), how many different outcomes can there be for (A, B)?

Solution

There are 6 possible rolls of the first dies, and 4 for the second, so $6 \cdot 4 = \boxed{24}$ possibilities for (A, B).

17. Principle of Indifference

If A is the roll of a fair six-sided die, (so $A \sim d6$) and B is the roll of the fair four-sided die (so $B \sim d4$), what would $\mathbb{P}((A, B) = (3, 1))$ using the Principle of Indifference?

Solution

There are $6 \cdot 4 = 24$ total outcomes, so if the Principle of Indifference applies, the result is $1/24$, which is $\boxed{0.04166\ldots}$.

Logical Operators

19. Logical AND, OR, and NOT

State if the following are true or false.

a. $(1 < 6) \vee (6 < 1)$

b. $(1 < 6) \wedge (6 < 1)$

a. $(6 < 1) \vee (1 < 6) \wedge (10 > 20)$

Solution

a. Since $(1 < 6)$ is true, that is enough to make $(1 < 6) \vee (6 < 1)$ equal to $\boxed{T}$.

b. Since $(6 < 1)$, is false, that is enough to make $(1 < 6) \wedge (6 < 1)$ equal to $\boxed{F}$.

c. By order of operations, the $\wedge$ goes first. Since $(10 > 20)$ is false, $(1 < 6) \wedge (10 > 20)$ is false as well. Combined with $(6 < 1) = F$, the whole statement is $\boxed{F}$.

21. Countable logical AND

Let x be a a positive integer (so $x \in \{1, 2, 3, \ldots\}$ is true.) State if $(x \geq 1) \wedge (x \geq 2) \wedge (x \geq 3) \wedge \cdots$ is true or false.

Solution

For any integer, eventually there will be a clause of the form $(x \geq x + 1)$ which is false. Since the countable logical AND contains at least one false value, the statement is $\boxed{F}$.

23. Logic to arithmetic

Write $\mathbb{I}(\neg s \vee \neg r)$ using only constants, $\mathbb{I}(s)$, and $\mathbb{I}(r)$.

Chapter 30: Encounters Resolved

Solution

Using our rules:

$$\mathbb{I}(\neg s \vee \neg r) = \mathbb{I}(\neg s) + \mathbb{I}(\neg r) - \mathbb{I}(\neg s)\mathbb{I}(\neg r)$$
$$= 1 - \mathbb{I}(s) + 1 - \mathbb{I}(r) -$$
$$(1 - \mathbb{I}(s))(1 - \mathbb{I}(r))$$
$$= 2 - \mathbb{I}(s) - \mathbb{I}(r) - 1 + \mathbb{I}(s) + \mathbb{I}(r) -$$
$$\mathbb{I}(s)\mathbb{I}(r)$$
$$= 1 - \mathbb{I}(s)\mathbb{I}(r).$$

25. Proof with Indicator Functions

Prove that $\neg(s \wedge r) = \neg s \vee \neg r$ using indicator functions.

Solution

The indicator function of the right hand side is

$$\mathbb{I}(\neg s \vee \neg r)$$
$$= \mathbb{I}(\neg s) + \mathbb{I}(\neg r) - \mathbb{I}(\neg s \neg r)$$
$$= 1 - \mathbb{I}(s) + 1 - \mathbb{I}(r) - \mathbb{I}(\neg s)\mathbb{I}(\neg r)$$
$$= 2 - \mathbb{I}(s) - \mathbb{I}(r) - (1 - \mathbb{I}(s))(1 - \mathbb{I}(r))$$
$$= 2 - \mathbb{I}(s) - \mathbb{I}(r) - 1 + \mathbb{I}(s) + \mathbb{I}(r) - \mathbb{I}(s)\mathbb{I}(r)$$
$$= 1 - \mathbb{I}(s)\mathbb{I}(r)$$
$$= 1 - \mathbb{I}(sr)$$
$$= \mathbb{I}(\neg(sr)),$$

which is the indicator function of the left hand side. Therefore, the associated logical expressions are equal. □

27. Logical Order of Operations

Write the order of operations for $s \vee w \wedge \neg r$ explicitly using parentheses.

Solution

Negatives come before logical AND which comes before logical OR. This ordering gives

$$\boxed{s \vee (w \wedge (\neg r))}.$$

29. De Morgan's Laws

Write $\neg(s_1 \wedge \neg s_2)$ using only logical OR and logical NOT.

Solution

Using De Morgan's laws and $\neg(\neg s_2) = s_2$ gives:

$$\boxed{\neg s_1 \vee s_2.}$$

31. Implication and Subsets

State whether the following are true or false.

a. $(x > 4) \rightarrow (x > 10)$.
b. $(x > 4) \rightarrow (x > 3)$.
c. $\{a, b, c\} \subseteq \{a, b, c, d\}$.
d. $\{a, b, c\} \subseteq \{a, c, e, g\}$.

Solution

a. This is $\boxed{F}$, for instance $x = 5$ has $(x > 4) = T$ and $(x > 10) = F$.
b. This is $\boxed{T}$. Since $(x > 4)$ and $(4 > 3)$, transitivity of the greater than symbol gives $(x > 3)$.
c. This is $\boxed{T}$. For $(x = a)$, $(x = b)$, and $(x = c)$, it holds that $x \in \{a, b, c, d\}$.
d. This is $\boxed{F}$ since $(b \in \{a, b, c\}) = T$ (so b is an element of the set on the left) but $(b \in \{a, c, e, g\}) = F$ (so b is not an element of the set on the right).

Rules of Probability

33. True Statements

For $X \in \mathbb{R}$, what is $\mathbb{P}(X^2 \geq 0)$?

Solution

If X is a real number, X^2 is always at least zero. Hence

$$\mathbb{P}(X^2 \geq 0) = \mathbb{P}(T) = 1,$$

so the answer is $\boxed{1}$.

35. False statements

For $Y \in \mathbb{R}$, what is $\mathbb{P}(|Y| < 0)$?

Solution
If Y is a real number, $|Y|$ is always at least zero. Hence
$$\mathbb{P}(|Y| < 0) = \mathbb{P}(\mathsf{F}) = 0,$$
so the answer is $\boxed{0}$.

37. Rolling the dice

Consider the following.
a. If $W \sim \text{d6}$, what is $\mathbb{P}(W \text{ is even})$?
b. If $R \sim \text{d10}$, what is $\mathbb{P}(R \leq 7)$?
c. If $Y \sim \text{d100}$, what is $\mathbb{P}(Y \leq 72)$?

Solution
a. The event that W is even is the same as $W \in \{2, 4, 6\}$, which is the disjoint logical OR of $W = 2$, $W = 4$, and $W = 6$. Each of these have probability $1/6$, so the answer is
$$a = \mathbb{P}(W \in \{2,4,6\})$$
$$= \mathbb{P}(W=2) + \mathbb{P}(W=4) + \mathbb{P}(W=6)$$
$$= \frac{1}{6} + \frac{1}{6} + \frac{1}{6} = \frac{1}{2},$$
or $\boxed{50\%}$.

b. There are 7 ways out of 10 that R can be at most 7, so $\boxed{70\%}$.

c. There are 72 ways out of 100 that Y can be at most 72, so $\boxed{72\%}$.

39. Using rules

If $\mathbb{P}(Y < 2) = 0.3$ and $\mathbb{P}(Y \in [2, 3]) = 0.4$, what is $\mathbb{P}(Y \in (-\infty, 3])$?

Solution
Note that $(Y < 2)$ and $(Y \in [2,3])$ are disjoint events. Also, $(Y < 2) \lor (Y \in [2,3]) = (Y \in (-\infty, 3]))$. Hence
$$\mathbb{P}(Y \in (-\infty, 3])) = \mathbb{P}(Y < 2) + \mathbb{P}(Y \in [2,3])$$
$$= 0.3 + 0.4,$$
which is $\boxed{70\%}$.

41. Negation rule

If $\mathbb{P}(G \geq 4) = 0.2$, what is $\mathbb{P}(G < 4)$?

Solution
Note $(G \geq 4) = \neg(G < 4)$. Hence
$$\mathbb{P}(G < 4) = 1 - \mathbb{P}(G \geq 4) = 1 - 0.2,$$
so the answer is $\boxed{80\%}$.

43. Implication

Given that $\mathbb{P}(R \leq 4) = 0.2$, what can be said about $\mathbb{P}(R \leq 5)$?

Solution
Since $(R \leq 4) \to (R \leq 5)$,
$$\boxed{\mathbb{P}(R \leq 5) \geq 20\%}.$$

45. Combining rules

Suppose $\mathbb{P}(X \geq 3) = 0.4$, $\mathbb{P}(X \leq 4) = 0.8$. What is $\mathbb{P}(X \in [3, 4])$?

Solution
Note that $\mathbb{P}(X \geq 3 \lor X \leq 4) = \mathbb{P}(\mathsf{T}) = 1$. Hence
$$1 = \mathbb{P}(X \geq 3) + \mathbb{P}(X \leq 4) - \mathbb{P}(X \geq 3, X \leq 4),$$
so $\mathbb{P}(X \in [3, 4]) = 0.4 + 0.8 - 1$ which is $\boxed{20\%}$.

Conditional Probability

47. A conditional die

Suppose $X \sim \text{d6}$. Find $\mathbb{P}(X = 1 \mid X \leq 4)$.

Solution
By the conditional probability formula:
$$\mathbb{P}(X = 1 \mid X \leq 4) = \frac{\mathbb{P}(X=1, X \leq 4)}{\mathbb{P}(X \leq 4)}$$
$$= \frac{\mathbb{P}(X=1)}{\mathbb{P}(X \leq 4)}$$
$$= \frac{1/6}{4/6} = 1/4,$$

which is 25%.

49. Interval Conditioning

The probability that $X \in [3, 7]$ is 0.423 and the probability that $X \leq 7$ is 0.620. What is $\mathbb{P}(X \in [3, 7] \mid X \leq 7)$?

Solution

By the conditional probability formula, this is

$$\mathbb{P}(X \in [3,7] \mid X \leq 7) = \frac{\mathbb{P}(X \in [3,7], X \leq 7)}{\mathbb{P}(X \leq 7)}$$
$$= \frac{\mathbb{P}(X \in [3,7])}{\mathbb{P}(X \leq 7)}$$
$$= \frac{0.423}{0.620},$$

which is approximately 0.6822.

51. Blood Tests

Given that a patient has a particular disease, the chance that a particular blood test comes back positive is 75%. The chance that the blood test comes back positive if the patient does not have the disease is 10%. The chance of having the disease is 3%. What is the chance that a patient gets back a positive?

Solution

Let p be the event that the blood test is positive, and d the event that the patient has the disease. Then the goal is to find $\mathbb{P}(p)$. The event p can be true either if d is true or d is false. That is,

$$\mathbb{P}(p) = \mathbb{P}(pd) + \mathbb{P}(p(\neg d)).$$

Using the two-stage rule:

$$\mathbb{P}(p) = \mathbb{P}(p \mid d)\mathbb{P}(d) + \mathbb{P}(p \mid \neg d)\mathbb{P}(\neg d).$$

Filling in our given numbers gives:

$$\mathbb{P}(p) = (75\%)(3\%) + (10\%)(97\%),$$

which is 11.95%.

CHAPTER 30: ENCOUNTERS RESOLVED

53. The Satellite

A weather satellite detects precipitation when precipitation exists 98% of the time. It gives a false positive and reports precipitation when none exists 4% of the time. If there is a 10% chance of precipitation, what is the chance that the satellite reports that there is precipitation?

Solution

Let p be the event of precipitation, and t the event of a positive test result. Then

$$\mathbb{P}(t) = \mathbb{P}(tp) + \mathbb{P}(t(\neg p))$$
$$= \mathbb{P}(p)\mathbb{P}(t \mid p) + \mathbb{P}(\neg p)\mathbb{P}(t \mid \neg p)$$
$$= (10\%)(98\%) + (90\%)(4\%),$$

which is 0.1340.

55. Raining Once More

If the odds that it will rain today are 4 to 5, what is the probability that it will rain today?

Solution

Let r be the event that it rains today. Then

$$\frac{\mathbb{P}(r)}{\mathbb{P}(\neg r)} = \frac{\mathbb{P}(r)}{1 - \mathbb{P}(r)} = \frac{4}{5},$$

so

$$5\mathbb{P}(r) = 4 - 4\mathbb{P}(r) \Rightarrow 9\mathbb{P}(r) = 4 \Rightarrow \mathbb{P}(r) = 4/9.$$

Hence the chance is $0.4444\ldots$.

57. Odds and Fairness

Prove that if the odds for s are greater than 1, then $\mathbb{P}(s) > 50\%$.

Solution

Suppose $\mathbb{P}(s)/\mathbb{P}(\neg s) > 1$. Then by the negation rule,

$$\frac{\mathbb{P}(s)}{1 - \mathbb{P}(s)} > 1,$$

which means
$$\mathbb{P}(s) > 1 - \mathbb{P}(s).$$
Bringing the $-\mathbb{P}(s)$ term over to the other side gives $2\mathbb{P}(s) > 1$, so $\mathbb{P}(s) > 1/2 = 50\%$ as desired. □.

59. Experience

Two out of seven staff members have experience working with R. If two staff members are chosen uniformly to be part of a task force, what is the chance that both will know R?

Solution

Let r_1 be the event that the first staff member chosen knows R, and r_2 the event that the second staff member chosen knows R. Then if the first staff member is chosen uniformly from the seven members,
$$\mathbb{P}(r_1) = \frac{2}{7}.$$

If r_1 is true, that leaves only 1 out of 6 staff members that know R. Hence
$$\mathbb{P}(r_2 \mid r_1) = \frac{1}{6}.$$

Combine to get
$$\mathbb{P}(r_1 r_2) = \mathbb{P}(r_1)\mathbb{P}(r_2 \mid r_1) = \frac{2}{7} \cdot \frac{1}{6},$$
which is about $\boxed{0.04761}$.

61. The Food Pantry

There are 9 cans in a food pantry that are unlabeled, but the inventory sheet says that they must be 4 cans of green beans and 5 cans of corn.
a. If 4 cans are chosen uniformly without replacement, and the first two cans chosen have green beans, what is that chance that the third can has green beans?
b. If 4 cans are chosen uniformly without replacement, what are the chances that all 4 have green beans?

Solution

Let s_i be the event that the ith chosen can has green beans.
a. The goal for this problem is to find
$$\mathbb{P}(s_3 \mid s_1 s_2).$$
If the first two cans have green beans, that leaves 2 cans with green beans out of 7 cans. So the desired probability is $2/7$, which is about $\boxed{0.2857}$.
b. Now the goal is to find $\mathbb{P}(s_1 s_2 s_3 s_4)$. Using conditioning:
$$\begin{aligned}\mathbb{P}(s_1 s_2 s_3 s_4) &= \mathbb{P}(s_1)\mathbb{P}(s_2 s_3 s_4 \mid s_1)\\ &= \frac{4}{9}\mathbb{P}(s_2 \mid s_1)\mathbb{P}(s_3 s_4 \mid s_1 s_2)\\ &= \frac{4}{9}\cdot\frac{3}{8}\mathbb{P}(s_3 \mid s_1 s_2)\mathbb{P}(s_4 \mid s_1 s_2 s_3)\\ &= \frac{4}{9}\cdot\frac{3}{8}\frac{2}{7}\frac{1}{6}.\end{aligned}$$

This is about $\boxed{0.007936}$.

Independence

63. Independence of Two Events

Suppose that $\mathbb{P}(a_1) = 0.3$ and $\mathbb{P}(a_2) = 0.6$, where a_1 and a_2 are independent events. What is $\mathbb{P}(a_1 a_2)$?

Solution

Since they are independent:
$$\mathbb{P}(a_1 a_2) = \mathbb{P}(a_1)\mathbb{P}(a_2) = (0.3)(0.6),$$
which is $\boxed{18\%}$.

65. All False

Suppose that s_1, s_2, s_3 are independent events each with probability 0.2.
a. What is the chance that none of the events occur?
b. What is the chance that at least one of the events occurs?

Solution

a. Since they are independent:

$$\mathbb{P}(\neg s_1 \neg s_2 \neg s_3) = \mathbb{P}(\neg s_1)\mathbb{P}(\neg s_2)\mathbb{P}(\neg s_3)$$
$$= (1 - 0.2)^3 = \boxed{0.5120}.$$

b. Note that "at least one of the events occurs" is the negation (by De Morgan's Laws) of the event "none of the events occur". So the answer is

$$1 - (1 - 0.2)^3 = \boxed{0.4880}.$$

67. The Archers

Four archers independently fire at a target. Each has a 0.2 chance of striking the target.
a. What is the chance that all the archers miss the target?
b. What is the chance that at least one archer hits the target?

Solution

a. If a_i is the event that archer i strikes the target, this this problem is to find

$$p = \mathbb{P}(\neg a_1 \wedge \neg a_2 \wedge \neg a_3 \wedge \neg a_4)$$
$$= \mathbb{P}(\neg a_1)\mathbb{P}(\neg a_2)\mathbb{P}(\neg a_3)\mathbb{P}(\neg a_4)$$
$$= (1 - \mathbb{P}(a_1)) \cdots (1 - \mathbb{P}(a_4))$$
$$= 0.8^4$$
$$= \boxed{0.4096}.$$

b. The negation of at least one archer hitting the target is no archers hitting the target, and that was found above.

$$b = \mathbb{P}(a_1 \vee \cdots \vee a_4)$$
$$= 1 - \mathbb{P}(\neg a_1 \cdots \neg a_4)$$
$$= 1 - 0.4096$$
$$= \boxed{0.5904}$$

69. Factory Woes

On a given day in a factory, there is a 3% chance of a shutdown, a 1% chance of a worker injury, and a 2% chance of a delivery delay. If these three events are independent, what is the chance that all three occur on a given day?

Solution

Since the events are independent, the probability that all three occur is just the product of the individual probabilities:

$$(0.03)(0.01)(0.02) = \boxed{0.000006000}.$$

Bayes Rule

71. Bayes' Rule

Suppose that $\mathbb{P}(X = 1) = 0.5$, $\mathbb{P}(X = 2) = 0.3$, $\mathbb{P}(X = 3) = 0.2$, and that $\mathbb{P}(r \mid X) = 1/X$.
a. What is $\mathbb{P}(r)$?
b. What is $\mathbb{P}(X = 1 \mid r)$?

Solution

a. Using the Law of Total Probability,

$$\mathbb{P}(r) = \sum_{i=1}^{3} \mathbb{P}(r \mid X = i)\mathbb{P}(X = i)$$
$$= (1/1)(0.5) + (1/2)(0.3) + (1/3)(0.2)$$
$$= 43/60,$$

which is about $\boxed{0.7166}$.

b. Using Bayes' Rule:

$$\mathbb{P}(X = 1 \mid r) = \mathbb{P}(r \mid X = 1) \frac{\mathbb{P}(X = 1)}{\mathbb{P}(r)}$$
$$= \frac{1}{1} \cdot \frac{1/2}{43/60},$$

which is $\boxed{0.6833}$.

73. Machine Problems

Machine A has a 1% chance of making a widget with an error, while Machine B has a 5% chance of error.
If Machine A makes 70% of the widgets and Machine B makes 30%, what is the chance that a widget with an error came from Machine A?

SOLUTION

Let a be the event that the widget came from Machine A, and e be the event that the widget had an error. Then from the problem

$$\mathbb{P}(a) = 0.7$$
$$\mathbb{P}(e \mid a) = 0.01$$
$$\mathbb{P}(e \mid \neg a) = 0.05.$$

Hence

$$\mathbb{P}(e) = \mathbb{P}(e \mid a)\mathbb{P}(a) + \mathbb{P}(e \mid \neg a)\mathbb{P}(\neg a)$$
$$= (0.01)(0.7) + (0.05)(0.3) = 0.022,$$

and by Bayes' Rule

$$\mathbb{P}(a \mid e) = \mathbb{P}(e \mid a)\frac{\mathbb{P}(a)}{\mathbb{P}(e)} = \frac{(0.01)(0.7)}{0.022},$$

which is $\boxed{0.3181}$.

75. CHOLESTEROL

Suppose that in a population, there is a 3% chance of having a genetic marker that doubles the chance of having high cholesterol. So if H is the event that the person has high cholesterol, and G is the event that they have the genetic marker, then

$$\mathbb{P}(H \mid G) = 2\mathbb{P}(H \mid G^C).$$

Given that someone has high cholesterol, what is the chance that they have the marker?

SOLUTION

Our goal is to find $\mathbb{P}(G \mid H)$. Using Bayes' Rule:

$$\mathbb{P}(G \mid H) = \frac{\mathbb{P}(H \mid G)\mathbb{P}(G)}{\mathbb{P}(H \mid G)\mathbb{P}(G) + \mathbb{P}(H \mid G^C)\mathbb{P}(G^C)}.$$

Replacing $\mathbb{P}(H \mid G^C)$ with $(1/2)\mathbb{P}(H \mid G)$ gives

$$\mathbb{P}(G \mid H) = \frac{\mathbb{P}(H \mid G)\mathbb{P}(G)}{\mathbb{P}(H \mid G)\mathbb{P}(G) + (1/2)\mathbb{P}(H \mid G)\mathbb{P}(G^C)}.$$

Note that $\mathbb{P}(H \mid G)$ cancels out of this expression!

$$\mathbb{P}(G \mid H) = \frac{0.03}{0.03 + (1/2)(0.97)},$$

which gives $\boxed{\mathbb{P}(G \mid H) = 0.05825}$.

77. A BIT OF A SCREW-UP

A company obtains screws from three manufacturers, code names X, Y, and Z. About 80% of screws come from company X, 10% from company Y, and 10% from company Z. Company X has about 2% of their screws defective, while company Y has only 1% defective, and company Z has 3% defective. Given that a particular screw is defective, find the probability that it came from company X, company Y, and company Z.

SOLUTION

Let d be the event that the screw is defective, and x, y, z be the events that the screw came from company X, Y, or Z respectively. Then the information in the problem can be summarized as

$$\mathbb{P}(x) = 0.8 \qquad \mathbb{P}(d|x) = 0.02$$
$$\mathbb{P}(y) = 0.1 \qquad \mathbb{P}(d|y) = 0.01$$
$$\mathbb{P}(z) = 0.1 \qquad \mathbb{P}(d|z) = 0.03.$$

The question is asking us to find

$$prob(x|d), \ \mathbb{P}(y|d), \ \mathbb{P}(z|d).$$

The answer can be found using Bayes' Rule, where

$$\mathbb{P}(x|d) = \mathbb{P}(d|x)\frac{\mathbb{P}(x)}{\mathbb{P}(d)}.$$

Both $\mathbb{P}(d|x)$ and $\mathbb{P}(x)$ are part of the problem statement, so the only thing needed is $\mathbb{P}(d)$. Fortunately (x, y, z)

CHAPTER 30: ENCOUNTERS RESOLVED

partition the truth, so

$$\begin{aligned}\mathbb{P}(d) &= \mathbb{P}(d(x \vee y \vee z)) \\ &= \mathbb{P}(dx) + \mathbb{P}(dy) + \mathbb{P}(dz) \\ &= \mathbb{P}(d|x)\mathbb{P}(x) + \mathbb{P}(d|y)\mathbb{P}(y) + \mathbb{P}(d|z)\mathbb{P}(z) \\ &= (0.02)(0.8) + (0.01)(0.1) + (0.03)(0.1) \\ &= 0.020.\end{aligned}$$

Hence

$$\mathbb{P}(x|y) = (0.02)\frac{0.8}{0.02} = 0.8.$$

Two more similar calculations give

$$\mathbb{P}(x|d) = 0.8000$$
$$\mathbb{P}(y|d) = 0.05000$$
$$\mathbb{P}(z|d) = 0.1500$$

So the chance it is from companies X, Y, and Z work out to be

$$\boxed{0.8000, 0.05000, 0, 1500}.$$

respectively.

RANDOM VARIABLES

79. TEN INDEPENDENT EVENTS

Let $X_1, \ldots, X_{10}$ be independent rolls of a fair six-sided die. What is the chance that no 5 is rolled?

SOLUTION

Let $r_i = (X_i = 5)$. Then our goal is to find:

$$p = \mathbb{P}(\neg r_1 \wedge \neg r_2 \wedge \cdots \neg r_{10}).$$

Since the events r_i are independent, so are $\neg r_i$, and

$$p = \prod_{i=1}^{10} \mathbb{P}(\neg r_i) = \prod_{i=1}^{10} 1 - \mathbb{P}(r_i) = \left(\frac{5}{6}\right)^{10},$$

which is about $\boxed{16.15\%}$.

81. FLIPS OF A 0-1 COIN

Suppose that X_1 and X_2 are random variables that are independent and equally likely to be 0 or 1.
a. If $X_1 = 1$, then what is the probability that $X_2 = 1$?
b. If $X_1 + X_2 \geq 1$ then what is the probability that $X_2 = 1$?

SOLUTION
a. Because they are independent,

$$\mathbb{P}(X_2 = 1 \mid X_1 = 1) = \mathbb{P}(X_2 = 1) = 1/2,$$

which is $\boxed{50\%}$.
b. Note that

$$(X_1 + X_2 \geq 1) = (X_1, X_2) \in \{(0,1), (1,0), (1,1)\}.$$

So

$$\begin{aligned}p &= \mathbb{P}(X_2 = 1 \mid X_1 + X_2 \geq 1) \\ &= \frac{\mathbb{P}(X_2 = 1, X_1 + X_2 \geq 1)}{\mathbb{P}(X_1 + X_2 \geq 1)} \\ &= \frac{1/2}{3/4} = \frac{2}{3},\end{aligned}$$

which is about $\boxed{66.66\%}$.

83. SIX ARROWS

An archer shoots six arrows at a target. Each hits (independently of the others) with probability 15%. What is the chance that at least five arrows hit?

SOLUTION
Let H be the number of arrows that hit the target. Then there is only one outcome $SSSSSS$ where all 6 arrows hit.

$$\mathbb{P}(SSSSSS) = (0.15)^6 = 0.0000139062\ldots.$$

There are six outcomes $FSSSSS, \ldots, SSSSSF$ where $H = 5$. Hence

$$\begin{aligned}\mathbb{P}(H = 5) &= 6\mathbb{P}(FSSSSS) \\ &= 6(0.15)^5(0.85)\end{aligned}$$

Therefore, the solution is

$$6(0.15)^5(0.85)+(0.15)^6 = (0.15)^5[6(0.85)+0.15],$$

or $\boxed{0.0003986\ldots}$.

85. COIN FLIPS

A bored merchant with a coin starts flipping the coin until the first head is seen. If G is the number of flips (including the final one) until a head is seen, then

$$\mathbb{P}(G = i) = (0.2)^{i-1}(0.8),$$

for all $i \in \{1, 2, \ldots\}$.

Verify that

$$\mathbb{P}(G \in \{1, 2, \ldots\}) = 1.$$

SOLUTION

The goal is to prove that $\mathbb{P}(G \in \{1, 2, \ldots\}) = 1$.

Proof. The events $(G = i)$ are disjoint for all $i \in \{1, 2, 3, \ldots\}$. Hence the countable additivity rule can be used to say:

$$\mathbb{P}(G \in \{1, 2, \ldots\}) = \mathbb{P}\left(\bigvee_{i=1}^{\infty}(G = i)\right)$$

$$= \sum_{i=1}^{\infty} \mathbb{P}(G = i)$$

$$= \sum_{i=1}^{\infty} (0.2)^{i-1}(0.8)$$

$$= \frac{0.8}{1 - 0.2} = 1,$$

as desired. □

87. MEASURABLE SETS

Suppose $[3, 4]$, $[4, 5)$, and $[5, 6)$ are measurable with respect to X. Prove that

$$\{[3, 5), [3, 6), [4, 6)\} \subseteq \mathcal{F}_X.$$

SOLUTION

Recall that the empty set is always in $\mathcal{F}$. Hence

$$[3, 4] \cup [4, 5) \cup \emptyset \cup \emptyset \cup \cdots = [3, 5) \in \mathcal{F}$$
$$[3, 4] \cup [4, 5) \cup [5, 6) \cup \emptyset \cup \emptyset \cup \cdots = [3, 6) \in \mathcal{F}$$
$$[4, 5) \cup [5, 6) \cup \emptyset \cup \emptyset \cup \cdots = [4, 6) \in \mathcal{F},$$

which completes the proof. □

89. THE TRIANGLE

Suppose that (X, Y) are equally likely to be $(0, 0)$, $(0, 1)$, or $(1, 0)$.

a. What is $\mathbb{P}(X = 0)$?
b. What is $\mathbb{P}(Y = 0)$?
c. What is $\mathbb{P}(X = 0, Y = 0)$?
d. Are X and Y independent?

SOLUTION

a. Each point has probability $1/3$ of occurring, and there are two points with $X = 0$, so

$$\mathbb{P}(X = 0) = 2/3 = \boxed{0.6666\ldots}.$$

b. Two out of three points has $Y = 0$, so

$$\mathbb{P}(Y = 0) = 2/3 = \boxed{0.6666\ldots}.$$

c. One point has both $X = 0$ and $Y = 0$, so

$$\mathbb{P}(X = 0, Y = 0) = 1/3 = \boxed{0.3333\ldots}.$$

d. The random variables X and Y are $\boxed{\text{not independent}}$, which we know because

$$\mathbb{P}(X = 0, Y = 0) = \frac{1}{3}$$

but

$$\mathbb{P}(X = 0)\mathbb{P}(Y = 0) = \frac{2}{3} \cdot \frac{2}{3} = \frac{4}{9} \neq \frac{1}{3} = \frac{3}{9}.$$

91. TRANSPORT

A bus arrives after a time given by the continuous random variable T.

a. What is $\mathbb{P}(T=4)$?
b. If $\mathbb{P}(T<4) = 0.4$, what is $\mathbb{P}(T \leq 4)$?
c. If $\mathbb{P}(T<4) = 0.4$, what is $\mathbb{P}(T > 4)$?

Solution
a. Because T is continuous, the probability it equals a particular value is $\boxed{0}$.

b. Because $(T<4)$ and $(T=4)$ are disjoint,

$$\mathbb{P}(T \leq 4) = \mathbb{P}(T<4) + \mathbb{P}(T=4) = \mathbb{P}(T<4).$$

So $\mathbb{P}(T \leq 4)$ is also $\boxed{0.4000}$.

c. Finally, $\mathbb{P}(T>4) = 1 - \mathbb{P}(T \leq 4)$, and from the last section of the problem, this is $\boxed{0.6000}$.

Uniform Random Variables

93. Summing Three Dice

Suppose $D_1 \sim$ d10, $D_2 \sim$ d20 and $D_3 \sim$ d6 are independent. What is

$$\mathbb{P}(D_1 + D_2 + D_3 = 36)?$$

Solution
The only way that the sum of these three variables can add to 36 is if $D_1 = 10$, $D_2 = 20$, and $D_3 = 6$. Therefore the answer is

$$\mathbb{P}(D_1 = 10, D_2 = 20, D_3 = 6) = \frac{1}{10} \cdot \frac{1}{20} \cdot \frac{1}{6} = \frac{1}{1200},$$

which is about $\boxed{0.0008333}$.

95. Standard Uniform

For $U \sim \text{Unif}([0,1])$, find:
a. $\mathbb{P}(U \leq 0.7)$.
b. $\mathbb{P}(U \leq -0.7)$.
c. $\mathbb{P}(U \leq 1.3)$.

Solution
a. Note that

$$\mathbb{P}(U \leq 0.7) = \mathbb{P}(U \in (-\infty, 0.7])$$
$$= \mathbb{P}(U \in (-\infty, 0.7] \cap [0,1]))$$
$$= \mathbb{P}(U \in [0, 0.7]).$$

So

$$\mathbb{P}(U \in (-\infty, 0.7]) = \frac{\text{Leb}([0, 0.7])}{\text{Leb}([0,1])}$$
$$= \frac{0.7 - 0}{1 - 0},$$

which is $\boxed{70\%}$.

b. Similarly,

$$\mathbb{P}(U \leq -0.7) = \mathbb{P}(U \in (-\infty, 0.7] \cap [0,1])$$
$$= \mathbb{P}(U \in \emptyset)$$
$$= \mathbb{P}(\mathsf{F}),$$

which is $\boxed{0}$.

c. Finally,

$$\mathbb{P}(U \leq 1.3) = \mathbb{P}(U \in (-\infty, 1.3] \cap [0,1])$$
$$= \mathbb{P}(U \in [0,1])$$
$$= \mathbb{P}(\mathsf{T}),$$

so the answer is $\boxed{1}$.

97. Functions of a Uniform

Prove for the discrete uniform $A \sim \text{Unif}(\{0,1,2\})$ that $A^2 \sim \text{Unif}(\{0,1,4\})$.

Solution
Note that

$$\mathbb{P}(A^2 = 0) = \mathbb{P}(A=0) = 1/3,$$
$$\mathbb{P}(A^2 = 1) = \mathbb{P}(A=1) = 1/3,$$
$$\mathbb{P}(A^2 = 4) = \mathbb{P}(A=2) = 1/3,$$

and so by definition,
$\boxed{A^2 \sim \text{Unif}(\{0,1,4\})}$.

99. Conditional Uniform

Suppose that $Y \sim$ d20. What is the distribution of:

a. Y given that $Y \geq 5$.
b. Y given that $Y < 5$.

SOLUTION
The distribution of a uniform conditioned on being in a set is also uniform over the conditioned set.
a. So $[Y \mid Y \geq 5] \sim \text{Unif}(\{5, 6, \ldots, 20\})$.
b. Here $[Y \mid Y < 5] \sim \text{Unif}(\{1, 2, 3, 4\})$.

101. DEFECTIVE TESTING

In a box with 20 parts, 2 are defective. An inspector picks a part out of the box uniformly at random. What is the probability that the inspector finds a defective part?

SOLUTION
Since two of the 20 parts are defective, there is a 2/20, or 10% chance of finding a defective part.

FUNCTIONS OF RANDOM VARIABLES

103. FINDING A CDF

If $X \sim \text{Unif}([-1, 2])$, find the cdf of $|X|$.

SOLUTION
Let $a < 0$. Then $\mathbb{P}(|X| \leq a) = 0$ since $|X|$ is always nonnegative.

Let $a > 2$. Then $\mathbb{P}(|X| \leq a) = 1$, since the absolute value of any number in $[-1, 2]$ is at most 2.

Now it gets interesting. Recall that $(|X| \leq a) = (-a \leq x \leq a)$ for $a \geq 0$. Let $a \in [0, 1]$. Then

$$\mathbb{P}(|X| \leq a) = \mathbb{P}(-a \leq X \leq a)$$
$$= \frac{a - (-a)}{3} = \frac{2}{3}a.$$

Let $a \in (1, 2]$. Then

$$\mathbb{P}(|X| \leq a) = \mathbb{P}(-a \leq X \leq a)$$
$$= \mathbb{P}(-1 \leq X \leq a)$$
$$= \frac{a - (-1)}{3}$$
$$= \frac{a + 1}{3}.$$

So the final result is

$$\frac{2}{3}a \mathbb{I}(a \in [0, 1]) + \frac{a + 1}{3}\mathbb{I}(a \in (1, 2]) + \mathbb{I}(a > 2).$$

This looks like this when plotted.

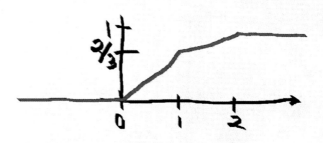

105. INDICATORS OF A UNIFORM

If $W \sim \text{Unif}([-2, 4])$ and $A = \mathbb{I}(W \leq 3)$, what is the distribution of A?

SOLUTION
Since A is either 0 or 1, it must have a Bernoulli distribution. The probability $A = 1$ is the probability $W \leq 3$, which is $(3 - (-2))/(4 - (-2)) = 5/6$. Hence $A \sim \text{Bern}(5/6)$.

107. SCALING AND SHIFTING UNIFORMS

Consider $U_1 \sim \text{Unif}([3, 8])$ and $U_2 \sim \text{Unif}([0, 1])$. Show that U_1 and $5U_2 + 3$ have the same cdf.

SOLUTION
First, the cdf of U_1 depends on whether or not $a < 3$, $a \in [3, 8]$, or $a > 8$. In the case $a < 3$, $\mathbb{P}(U_1 \leq a) = 0$. In the case

$a > 8$, $\mathbb{P}(U_1 \leq a) = 1$. When $a \in [3, 8]$, the probability for uniforms gives

$$\mathbb{P}(U_1 \leq a) = \frac{a-3}{8-3} = \frac{a-3}{5}.$$

Combining these cases gives

$$\text{cdf}_{U_1}(a) = \frac{a-3}{5}\mathbb{I}(a \in [3,8]) + \mathbb{I}(a > 8).$$

Similarly, the cdf of U_2 is

$$\text{cdf}_{U_2}(a) = a\mathbb{I}(a \in [0,1]) + \mathbb{I}(a > 1).$$

Since $5U_2 + 3 \leq a$ if and only if $U_2 \leq (a-3)/5$,

$$\text{cdf}_{5U_2+3}(a)$$
$$= \text{cdf}_{U_2}((a-3)/5)$$
$$= \frac{a-3}{5}\mathbb{I}\left(\frac{a-3}{5} \in [0,1]\right) + \mathbb{I}\left(\frac{a-3}{5} > 1\right)$$
$$= \frac{a-3}{5}\mathbb{I}(a \in [3,8]) + \mathbb{I}(a > 8).$$

Therefore, both U_1 and $5U_2 + 3$ have the same cdf!

109. Scaling Exponentials

Recall that if $U \sim \text{Unif}([0,1])$, $-\ln(U)/\lambda \sim \text{Exp}(\lambda)$. Use this to prove that if $X \sim \text{Exp}(\lambda)$, $X/c \sim \text{Exp}(c\lambda)$ for any nonnegative constant c.

Solution
Let $X \sim \text{Exp}(\lambda)$. Then $X = -\ln(U)/\lambda$ where $U \sim \text{Unif}([0,1])$. Hence $X/c = -\ln(U)/[c\lambda]$, and $X/c \sim \text{Exp}(c\lambda)$. □

111. Conditional Expectations

Suppose $T \sim \text{Exp}(2.4)$. What is $\mathbb{P}(T \geq 4 \mid T \geq 1)$?

Solution
Recall $T = -(1/2.4)\ln(U)$. So

$$(T \geq t) = (-(1/2.4)\ln(U) \geq t)$$
$$= (\ln(U) \leq -2.4t)$$
$$= (U \leq \exp(-2.4t)).$$

Hence

$$\mathbb{P}(T \geq 4 \mid T \geq 1) = \frac{\mathbb{P}(T \geq 4, T \geq 1)}{\mathbb{P}(T \geq 1)}$$
$$= \frac{\mathbb{P}(T \geq 4)}{\mathbb{P}(T \geq 1)}$$
$$= \frac{\exp(-2.4 \cdot 4)}{\exp(-2.4 \cdot 1)}$$
$$= \exp(-2.4 \cdot 3),$$

which is about $\boxed{0.0007465}$.

113. The Ceiling Function

Let $\lceil x \rceil$ be the ceiling function that is the smallest integer greater than or equal to x. So $\lceil 4.3 \rceil = \lceil 5 \rceil = 5$. Note that for an integer i, $\lceil x \rceil = i$ if and only if $i - 1 < x \leq i$. For $U \sim \text{Unif}([0,1])$ find
a. $\mathbb{P}(\lceil 2U \rceil = 2)$
b. $\mathbb{P}(\lceil 2U \rceil = 1)$
c. $\mathbb{P}(\lceil 2U \rceil = 0)$

Solution
a. Note

$$\mathbb{P}(\lceil 2U \rceil = 2) = \mathbb{P}(1 < 2U \leq 2)$$
$$= \mathbb{P}(1/2 < U \leq 1)$$
$$= (1 - 1/2)/(1 - 0),$$

which is $\boxed{50\%}$.
b. Similarly, this is

$$\mathbb{P}(\lceil 2U \rceil = 1) = \mathbb{P}(0 < 2U \leq 1)$$
$$= \mathbb{P}(0 < U \leq 1/2)$$
$$= (1/2 - 0)/(1 - 0),$$

which is $\boxed{50\%}$.

c. Finally, this is

$$\mathbb{P}(\lceil 2U \rceil = 0) = \mathbb{P}(-1 < 2U \leq 0)$$
$$= \mathbb{P}(-1/2 < U \leq 0)$$
$$= 0.$$

There is no chance of this happening, so the answer is $\boxed{0}$.

THE BERNOULLI PROCESS

115. BASIC BINOMIALS

Suppose that $X \sim \text{Bin}(100, 0.05)$.
a. What is $\mathbb{P}(X = 0)$?
b. What is $\mathbb{P}(X = 1)$?

SOLUTION
a. This is $\mathbb{P}(FFF \cdots F) = (0.95)^{100}$ which is about $\boxed{0.005920}$.
b. This is

$$\binom{100}{1}(0.05)(0.95)^{99} = \frac{100!}{99!1!} \cdot (0.05)(0.95)^{99}$$
$$= 100 \cdot (0.05)(0.95)^{99},$$

which is about $\boxed{0.03116}$.

117. BASIC GEOMETRICS

Let $G \sim \text{Geo}(0.3)$. Find $\mathbb{P}(G \geq 3)$.

SOLUTION
Recall $G = \inf\{i : B_i = 1\}$ where $B_1, B_2, \ldots$ are iid $\text{Bern}(0.3)$. Note that $G \geq 3$ if and only if $B_1 = B_2 = 0$. The probability this happens is

$$\mathbb{P}(B_1 = B_2 = 0) = \mathbb{P}(B_1 = 0)\mathbb{P}(B_2 = 0)$$
$$= (0.7)(0.7),$$

so the result is $\boxed{49\%}$.

119. NEGATIVE BINOMIALS

Suppose that $R \sim \text{NegBin}(4, 0.25)$.
a. Find $\mathbb{P}(R = 10)$.
b. Find $\mathbb{P}(R \leq 2)$.

SOLUTION
a. This is

$$\binom{9}{3}0.25^4 0.75^{10-4} = \boxed{0.05839}$$

b. This is $\boxed{0}$ because you need at least as 4 trials before you get 4 successes so $\mathbb{P}(R \leq 2) = 0$.

121. THE DRUG TRIAL

A drug trial independently brings in 18 patients. Each patient is given a drug expected to lower blood pressure. If the probability the drug works for any given person is 20%, what is the chance that the drug works for at most 5 patients?

SOLUTION

$$\mathbb{P}(N \leq 5) = \sum_{i=0}^{5} \mathbb{P}(N = i).$$

Here

$$\mathbb{P}(N = 0) = \binom{18}{0}0.2^0 0.8^{18} = 0.0180144\ldots$$

$$\mathbb{P}(N = 1) = \binom{18}{1}0.2^1 0.8^{17} = 0.08106479\ldots$$

$$\mathbb{P}(N = 2) = \binom{18}{2}0.2^2 0.8^{16} = 0.17226269\ldots$$

$$\mathbb{P}(N = 3) = \binom{18}{3}0.2^3 0.8^{15} = 0.22968358\ldots$$

$$\mathbb{P}(N = 4) = \binom{18}{4}0.2^4 0.8^{14} = 0.21532836\ldots$$

$$\mathbb{P}(N = 5) = \binom{18}{5}0.2^5 0.8^{13} = 0.15072985\ldots$$

so

$$\mathbb{P}(N \leq 5) \approx \boxed{0.8670}.$$

123. THE BAD BATCH

A company makes about 5,000 of a particular part a year. If there is a 0.001 chance (independently) that each part is a failure, find the probability that there are no failures among the parts.

SOLUTION
Let $N \sim \text{Bin}(5000, 0.001)$, then the chance of no failures is $\mathbb{P}(N = 0)$

$$\binom{5000}{0} 0.001^0 0.999^{5000} \approx \boxed{0.006721}$$

125. Fly away

A plane scout looking for forest fires during July in Montana has a 3% chance (independently) of noticing a fire each time a flight is taken. What is the chance that more than 30 flights are needed before the first fire is seen?

SOLUTION
In order to require more than 30 flights, the first 30 flights must be failures. This happens (because of independence) with probability

$$(1 - 0.03)^{30} = \boxed{0.4010\ldots}.$$

THE POISSON POINT PROCESS

127. Tails of exponentials

If A is a standard exponential, what is $\mathbb{P}(A \geq 1)$?

SOLUTION
Since $A = \ln(1/U)$ where $U \sim \text{Unif}([0, 1])$,

$$\begin{aligned}\mathbb{P}(A \geq 1) &= \mathbb{P}(\ln(1/U) \geq 1) \\ &= \mathbb{P}(1/U \geq \exp(1)) \\ &= \mathbb{P}(U \leq \exp(-1)) \\ &= \exp(-1),\end{aligned}$$

which is about $\boxed{36.78\%}$.

129. Bernoullis from geometrics

If $G_1 = 2$, $G_2 = 2$, and $G_3 = 5$ are used to create a Bernoulli point process, what are the values of $B_1, \ldots, B_7$?

SOLUTION
The point process will be

$$\{2, 2 + 2, 2 + 2 + 5, \ldots\} = \{2, 4, 9, \ldots\}.$$

Hence $B_2 = B_4 = B_9 = 1$, but $B_i = 0$ for $i \notin \{2, 4, 9, \ldots\}$. Hence

$$\boxed{(B_1, \ldots, B_7) = (0, 1, 0, 1, 0, 0, 0)}.$$

131. Understanding Poisson point processes

Let $P = \{T_1, T_2, \ldots\}$ where $T_1 < T_2 < \cdots$ be a Poisson point process of rate 1.2.
a. What is the distribution of T_1?
b. What is the distribution of T_2?
c. What is the probability that there are exactly two points in $[0, 1]$?

SOLUTION
a. This is $\boxed{\text{Exp}(1.2)}$.
b. This is $\boxed{\text{Gamma}(2, 1.2)}$.
c. Since $N_{[0,1]} \sim \text{Pois}(1.2 \cdot (1 - 0))$,

$$\mathbb{P}(N_{[0,1]}) = 2) = \exp(-1.2)\frac{1.2^2}{2!},$$

which is about $\boxed{0.2168}$.

133. Conditional PPP

Let $P \sim \text{PPP}(1.4)$. What is the probability that there are two points in $[0, 2]$, given that there are four points in $[0, 4]$?

SOLUTION
Note $N_{[0,4]} = N_{[0,2]} + N_{[2,4]}$ (Note $N_{[2,2]} = 0$ with probability 1.) If $N_{[0,4]} = 4$ and $N_{[0,2]} = 2$, then $N_{[2,4]} = 4 - 2 = 2$ as well.

So the answer a that is the target is
$$\begin{aligned}a &= \mathbb{P}(N_{[0,2]} = 2 \mid N_{[0,4]=4})\\ &= \frac{\mathbb{P}(N_{[0,2]} = 2, N_{[0,4]} = 4)}{\mathbb{P}(N_{[0,4]} = 4)}\\ &= \frac{\mathbb{P}(N_{[0,2]} = 2, N_{[2,4]} = 2)}{\mathbb{P}(N_{[0,4]} = 4)}\\ &= \frac{\mathbb{P}(N_{[0,2]} = 2)\mathbb{P}(N_{[2,4]} = 2)}{\mathbb{P}(N_{[0,4]} = 4)}\end{aligned}$$

135. THE RESTAURANT

A restaurant models arriving customers as a Poisson point process of rate 70 per hour.

What is the chance that there is at least one customer arrival in the first minute?

SOLUTION
First convert everything to minutes. One hour equals 60 minutes, so
$$\frac{70}{\text{hour}} = \frac{70}{60 \text{ min}} = \frac{7}{6} \text{ per min}$$
Hence
$$N_{[0,1]} \sim \text{Pois}((7/6)(1-0)),$$
and the chance this is at least 1 is
$$\begin{aligned}\mathbb{P}(N_{[0,1]} \geq 1) &= 1 - \mathbb{P}(N_{[0,1]} = 0)\\ &= 1 - \exp(-7/6)\frac{(7/6)^0}{0!},\end{aligned}$$
or about $\boxed{0.6885}$.

DENSITIES FOR CONTINUOUS RANDOM VARIABLES

137. DENSITY OF A UNIFORM

Suppose $\mathbb{P}(Y \in dy) = (1/30)\mathbb{I}(y \in [0,30])\,dy$. What is $\mathbb{P}(Y \leq 5)$?

SOLUTION
This is
$$\begin{aligned}\mathbb{P}(Y \leq 5) &= \int_{y\in(-\infty,5]} \mathbb{P}(Y \in dy)\\ &= \int_{y\in(-\infty,5]} (1/30)\mathbb{I}(y \in [0,30])\,dy\\ &= \int_{y\in[0,5]} (1/30)\,dy\\ &= (1/30)(5-0) = 1/6,\end{aligned}$$
which is about $\boxed{0.1666}$.

139. MORE UNIFORM DENSITY

Say that
$\mathbb{P}(W \in dw) = (1/20)\mathbb{I}(w \in [30,50])\,dw$.
What is the density of W?

SOLUTION
The density is just $\mathbb{P}(W \in dw)/dw$, or
$$\boxed{(1/20)\mathbb{I}(w \in [30,50]).}$$

141. ANOTHER EXPONENTIAL DENSITY

Suppose $\mathbb{P}(R \in dr) = 3\exp(-3r)\mathbb{I}(r \geq 0)\,dr$. Find $\text{cdf}_R(a)$.

SOLUTION
For $r < 0$,
$$\begin{aligned}\mathbb{P}(R \leq r) &= \int_{-\infty}^r 3\exp(-3r)\mathbb{I}(r \geq 0)\,dr\\ &= \int_{-\infty}^r 0\,dr = 0,\end{aligned}$$
for $r \geq 0$,
$$\begin{aligned}\mathbb{P}(R \leq r) &= \int_{-\infty}^r 3\exp(-3r)\mathbb{I}(r \geq 0)\,dr\\ &= \int_0^r 3\exp(-3r)\,dr\\ &= -\exp(-3r)|_0^r\\ &= 1 - \exp(-3r).\end{aligned}$$

Therefore the cdf is

$$\boxed{\mathrm{cdf}_R(r) = [1 - \exp(-3r)]\mathbb{I}(r \geq 0).}$$

143. Normalizing an Exponential

Suppose $\mathbb{P}(R \in dr) = C\exp(-3r)\mathbb{I}(r \in [0,2])\,dr$. Find C.

Solution
Integrate the density from $-\infty$ to ∞,

$$\int_{-\infty}^{\infty} C\exp(-3r)\mathbb{I}(r \in [0,2])\,dr = 1$$

$$\int_0^2 C\exp(-3r)\,dr = 1$$

$$C(-1/3)\exp(-3r)\big|_0^2 = 1$$

$$C(1/3)[1 - \exp(-6)] = 1$$

$$3/[1 - \exp(-6)] = C,$$

which gives about $\boxed{C = 3.007}$.

145. Density of a Function of a Uniform

Suppose $U \sim \mathsf{Unif}([0,1])$, and $W = \sqrt{U}$.
a. Find cdf_W.
b. Find pdf_W.

Solution
Use the function relationship.
a. Let $w \in \mathbb{R}$:

$$\mathbb{P}(W \leq w) = \mathbb{P}(\sqrt{U} \leq w)$$

which is 0 if $w \leq 0$. If $w \geq 0$,

$$\mathbb{P}(W \leq w) = \mathbb{P}(U \leq w^2).$$

This is just w^2 if $w \in [0,1]$, and 1 if $w > 1$. Hence the cdf is

$$\boxed{\mathrm{cdf}_W(w) = w^2\mathbb{I}(w \in [0,1]) + \mathbb{I}(w > 1).}$$

b. To find the pdf, differentiate:

$$\begin{aligned}\mathrm{pdf}_W(w) &= [w^2\mathbb{I}(w \in [0,1]) + \mathbb{I}(w > 1)]' \\ &= [w^2\mathbb{I}(w \in [0,1])]' + [\mathbb{I}(w > 1)]' \\ &= \mathbb{I}(w \in [0,1])[w^2]' + \mathbb{I}(w > 1)[1]' \\ &= 2w\mathbb{I}(w \in [0,1]).\end{aligned}$$

That is,

$$\boxed{\mathrm{pdf}_W(w) = 2w\mathbb{I}(w \in [0,1]).}$$

147. From CDF to PDF

Suppose $\mathrm{cdf}_X(x) = (1 - 1/x)\mathbb{I}(x \geq 1)$. Find $\mathrm{pdf}_X(x)$.

Solution
Differentiate to get:

$$\boxed{\mathrm{pdf}_X(x) = (1/x^2)\mathbb{I}(x \geq 1).}$$

Find $\mathrm{pdf}_X(x)$.

Densities for Discrete Random Variables

149. PDF for a Die

If $Y \sim \mathsf{d}10$, what is $\mathrm{pdf}_Y(i)$?

Solution
This is just $\mathbb{P}(Y = i)$, which is

$$\boxed{\mathbb{P}(Y = i) = \frac{1}{10}\mathbb{I}(i \in \{1,\ldots,10\}).}$$

151. Binomial Mode

Suppose $\mathrm{pdf}_B(i) = \binom{10}{i}p^i(1-p)^{10-i}$.
a. What is $\mathbb{P}(B = 6)$?
b. What is the mode if $p = 0.42$?

SOLUTION

a. This probability is

$$\mathbb{P}(B = 6) = \mathrm{pdf}_B(6) = \frac{10!}{6!4!}p^6(1-p)^4,$$

or $\boxed{210p^6(1-p)^4}$.

b. Note that for $i \in \{0, \ldots, 9\}$,

$$\frac{\mathrm{pdf}_B(i+1)}{\mathrm{pdf}_B(i)}$$

is the product

$$\frac{10!/[(i+1)!(10-i-1)!]}{10!/[i!(10-i)!]} \frac{p^{i+1}(1-p)^{10-i-1}}{p^i(1-p)^{10-i}}.$$

Canceling as much as possible gives

$$\frac{\mathrm{pdf}_B(i+1)}{\mathrm{pdf}_B(i)} = \frac{9-i}{(i+1)} \cdot \frac{p}{1-p}.$$

When $p = 0.42$, this is

$$\frac{9-i}{i+1} \cdot \frac{0.42}{0.58}.$$

When $i = 0$ this is much greater than 1, as i increases it starts going down. Solving for when this is strictly less than 1 gives $i > 16/5 = 3.2$. Hence the mode is at $\boxed{i = 4}$.

153. GAMMA MEDIAN

Suppose $\mathrm{pdf}_G(r) = (1/6)r^3 \exp(-r)\mathbb{I}(r \geq 0)$. Find the median of G.

SOLUTION
Note for $a \geq 0$,

$$\mathbb{P}(G \leq a) = \int_{-\infty}^{a} (1/6)r^3 \exp(-r)\mathbb{I}(r \geq 0) \, dr$$
$$= \int_0^a (1/6)r^3 \exp(-r)\mathbb{I}(r \geq 0) \, dr$$
$$= 1 - (1 + a + (1/2)a^2 + (1/6)a^3)\exp(-a).$$

Setting that equal to $1/2$ and solving gives a about $\boxed{3.672}$.

155. CDF OF A SCALED DIE

For $X \sim d6$, what is the cdf of $2X$?

SOLUTION
Saying $X \sim d6$ is the same as saying $X \sim \mathrm{Unif}(\{1,2,3,4,5,6\})$. That makes $2X \sim \mathrm{Unif}(\{2,4,6,8,10,12\})$. Therefore, there is a jump of size $1/6$ at each of $\{2,4,6,8,10,12\}$ in the cdf.

So one way to write the cdf is with indicator functions:

$$\mathrm{cdf}_X(a) = \frac{1}{6}[\mathbb{I}(a \geq 2) + \mathbb{I}(a \geq 4) + \cdots + \mathbb{I}(a \geq 12)].$$

Another way is with the floor function. The floor function (written $\mathrm{floor}(x)$ or $\lfloor x \rfloor$) rounds x down to the next integer. For instance, $\lfloor 2.3 \rfloor = \lfloor 2 \rfloor = 2$.

For the cdf at a, the goal is to multiply $1/6$ by the number of integer multiples of 2 given, as long as a is between 2 and 12. For a greater than 12, the cdf is just 1. So this can be done with

$$\boxed{\mathrm{cdf}_X(a) = \frac{1}{6}\lfloor a/2 \rfloor \mathbb{I}(2 \leq a \leq 12) + \mathbb{I}(a > 12).}$$

157. SCALING A BETA

For T with $\mathrm{pdf}_T(t) = 12t^2(1-t)\mathbb{I}(t \in [0,1])$, what is the pdf of $2T$?

SOLUTION
Use the shifting and scaling formula for continuous random variables.

$$f_{2T}(t) = \frac{1}{2}f_T(t/2)$$
$$= (12/2)(1/4)t^2(1 - t/2)\mathbb{I}(t/2 \in [0,1]),$$

and simplifying gives

$$\boxed{f_{2T}(t) = (3/2)t^2(1 - t/2)\mathbb{I}(t \in [0,2]).}$$

159. SURVIVAL FUNCTION OF AN EXPONENTIAL

For $T \sim 4\exp(-4t)\mathbb{I}(t \geq 0)$, what is the survival function of T?

SOLUTION
The survival function is
$$\mathrm{sur}_T(a) = 1 - \mathrm{cdf}_T(a).$$

The cdf of an exponential is
$$0 \cdot \mathbb{I}(a < 0) + (1 - \exp(-4a))\mathbb{I}(a \geq 0),$$

so the survival function is

$$\boxed{\mathrm{sur}_T(a) = \mathbb{I}(a < 0) + \exp(-4a)\mathbb{I}(a \geq 0).}$$

MEAN OF A RANDOM VARIABLE

161. MEAN OF FINITE RANDOM VARIABLES

If $\mathbb{P}(W = 1) = \mathbb{P}(W = 2) = 0.13$, and $\mathbb{P}(W = 4) = 0.74$, what is $\mathbb{E}[W]$?

SOLUTION
This will be
$$(1)(0.13) + (2)(0.13) + (4)(0.74),$$
which is $\boxed{3.350}$.

163. MEAN OF A DISCRETE UNIFORM

Suppose $U \sim \mathrm{Unif}(\{0, 10, 100\})$. What is $\mathbb{E}[U]$?

SOLUTION
Since U is uniform over three values, each has a $1/3$ chance of being the output. Hence
$$\mathbb{E}[U] = \frac{1}{3}[0 + 10 + 100] = \frac{110}{3},$$
which is about $\boxed{36.66}$.

165. MEAN OF A DIE ROLL

Suppose $X \sim \mathrm{d}8$. What is $\mathbb{E}[\mathbb{I}(X \leq 3)]$?

SOLUTION
The mean of an indicator function is just the probability of the event in the indicator function. Here, that is $3/8$, or $\boxed{0.3750}$.

167. MEAN OF A DIFFERENCE

Suppose $\mathbb{E}[A] = 1.2$ and $\mathbb{E}[B] = 6.3$. What is $\mathbb{E}[A - B]$?

SOLUTION
By linearity, this is
$$\mathbb{E}[A - B] = \mathbb{E}[A] - \mathbb{E}[B],$$
which is $1.2 - 6.3$, or $\boxed{-5.100}$.

169. SYMMETRY OF A DISCRETE UNIFORM

Suppose W is uniform over $\{-5, 0, 5\}$.
a. Show that W is symmetric around 0.
b. What is the expected value of W?

SOLUTION
a. Note W is uniform over $\{-5, 0, 5\}$, so $-W$ is uniform over $\{5, 0, -5\}$. But this is the same set, so W and $-W$ have the same distribution!
b. Since W is symmetric around 0 (and has an expected value), $\boxed{\mathbb{E}[W] = 0}$.

171. VERTIGON'S ARMY

The Dark Lord Vertigon was believed to have (with equal probability) a thousand, six thousand, or eight thousand soldiers in his army. What was the expected size of Vertigon's Army?

SOLUTION
Multiply each outcome times the probability of that outcome and sum to get
$$\mathbb{E}[A] = (1/3)(1000) + (1/3)(6000) + (1/3)(8000)$$
$$= 15000/3 = \boxed{5000}.$$

173. Four stores

Pretty Polly's Pet Store has four locations. The first averages 200 customers a day, the second averages 232, the third 330, and the last 280. Altogether, what is the total average number of customers at all of the four stores per day?

Solution
This will be

$$\mathbb{E}[N] = \frac{1}{4}(200) + \frac{1}{4}(232) + \frac{1}{4}(330) + \frac{1}{4}(280)$$
$$= \boxed{260.5}.$$

175. The SLLN in action

Suppose that W has mean 2 and $W_1, W_2, \ldots$ are iid W. What can be said about

$$\lim_{n \to \infty} \frac{W_1 + \cdots + W_n}{n}?$$

Solution
By the Strong Law of Large numbers, this is equal to $\mathbb{E}[W] = \boxed{2}$ with probability 1.

177. Two tasks

Suppose a certain task can be broken down into two parts. Part 1 requires T_1 amount of time and Part 2 requires T_2 amount of time. Part 2 cannot be started until Part 1 is complete. If $\mathbb{E}[T_1] = 4.2$ hours and $\mathbb{E}[T_2] = 1.3$ hours, what is the average time needed to complete both tasks?

Solution
This is just

$$\mathbb{E}[T_1 + T_2] = \mathbb{E}[T_1] + \mathbb{E}[T_2] = 4.2 + 1.3,$$

or $\boxed{5.500 \text{ hours}}$.

Mean of a General Random Variable

179. Mean of an Exponential

For $W \sim \text{Exp}(-2)$, set up the following integrals.
a. $\mathbb{E}[W]$.
b. $\mathbb{E}[W^2]$.
c. $\mathbb{E}[\mathbb{I}(W < 3)]$

Solution
The density is $f_W(w) = 2\exp(-2w)\mathbb{I}(w \geq 0)$. Inside the integral, replace the random variable with the index variable.

a. $\boxed{\mathbb{E}[W] = \int 2w \exp(-2w)\mathbb{I}(w \geq 0) \, dw}$

b. $\boxed{\mathbb{E}[W^2] = \int 2w^2 \exp(-2w)\mathbb{I}(w \geq 0) \, dw}$

c.
$\boxed{\mathbb{E}[\mathbb{I}(W < 3)] = \int 2\exp(-2w)\mathbb{I}(0 \leq w < 3) \, dw}$

181. Mean of functions of a continuous uniform

For a random variable $T \sim \text{Exp}(\lambda)$, so $f_T(t) = \lambda \exp(-\lambda t)\mathbb{I}(t \geq 0)$, find $\mathbb{E}[T]$.

Solution
This will be

$$\mathbb{E}[T] = \int t\lambda \exp(-\lambda t)\mathbb{I}(t \geq 0) \, dt$$
$$= \int_0^\infty t[-\exp(-\lambda t)]' \, dt$$
$$= (t)(-\exp(-\lambda t))|_0^\infty - \int_0^\infty [t]'(-\exp(-\lambda t)) \, dt$$
$$= 0 - 0 - \frac{\exp(-\lambda t)}{\lambda}\bigg|_0^\infty$$
$$= \frac{1}{\lambda}.$$

Hence the result is $\boxed{1/\lambda}$.

CHAPTER 30: ENCOUNTERS RESOLVED

183. Monte Carlo with Uniforms

Using the function `runif` that generates $U \sim \text{Unif}([0,1])$, write R code to estimate

$$\int_0^1 \sqrt{u}\, du$$

using 10^6 samples.

Solution
In R, use

```
u <- runif(10^6)
print(mean(u^(1 / 2)))
```

The result will be close to $\boxed{0.6666186}$. Of course, every time you run this code, you will get a slightly different answer, because new random uniforms are being generated each time!

185. Polynomials of Continuous Uniforms

Let $U \sim \text{Unif}([0,1])$. Find $\mathbb{E}[(1-U)(1+U)]$.

Solution
This is

$$\mathbb{E}[(1-U)(1+U)] = \mathbb{E}[1-U^2]$$
$$= \int (1-u^2)\mathbb{I}(u \in [0,1])\, du$$
$$= \int_0^1 (1-u^2)\, du$$
$$= (u - u^3/3)\big|_0^1$$
$$= 2/3,$$

which is about $\boxed{0.6666}$.

187. A Nonintegrable Density

Suppose X has density $f_X(x) = (4/\tau)/(1+x^2)$. Show that X is not integrable.

Solution
Consider finding $\mathbb{E}[X]$ using the integral

$$I = \int_{-\infty}^{\infty} x \frac{4/\tau}{1+x^2}\, dx$$
$$= 2\int_0^{\infty} x \frac{4/\tau}{1+x^2}\, dx$$
$$= 2\int_0^{\infty} \frac{4}{\tau} \cdot \frac{1}{2}[\ln(1+x^2)]'\, dx$$
$$= \frac{2}{\tau}\left[\lim_{b \to \infty} \ln(1+b^2) - 0\right]$$
$$= \infty.$$

This is not a finite value, and so the random variable X is not integrable.

189. Deriving Formulas

Suppose $\mathbb{E}[X] = \mu$. Prove that

$$\mathbb{E}[(X-\mu)^2] = \mathbb{E}[X^2] - \mu^2.$$

Solution
Proof. Expanding the square, and using linearity of expectations,

$$\mathbb{E}[(X-\mu)^2] = \mathbb{E}[X^2 - 2X\mu + \mu^2]$$
$$= \mathbb{E}[X^2] - 2\mathbb{E}[X]\mu + \mu^2$$
$$= \mathbb{E}[X^2] - 2\mu^2 + \mu^2$$
$$= \mathbb{E}[X^2] - \mu^2,$$

completing the proof. □

Conditional Expectation

191. Raising Sales

A marketing firm believes that an ad campaign has a 30% chance of raising sales exactly 100%, a 50% chance of raising sales exactly 20%, and a 20% chance of having no effect on sales.

a. Calculate the expected raise in sales.

b. Now suppose that instead of exactly raising sales 100%, 20% and 0%, there is a 30% chance of raising sales a random amount that has average value 100%, a 50% chance of raising sales a random amount that has average 20%, and a 20% chance of raising sales a random amount that has average 0%. Now calculate the expected raise in sales.

Solution
a. Let S denote the raise in sales. Then

$$\mathbb{P}(S=1)=0.3,\ \mathbb{P}(S=0.2)=0.5,\ \mathbb{P}(S=0)=0.2$$

So

$$\mathbb{E}[S] = (0.3)(1)+(0.2)(0.5)+(0)(0.2) = \boxed{0.4000}.$$

b. Let $X=1$ if there is a 100% average rise in sales, $X=2$ if there is a 20% average rise in sales, and $X3$ if there is a 0% average rise in sales. Then

$$\mathbb{E}[S|X=1] = 1 \qquad \mathbb{P}(X=1) = 0.3$$
$$\mathbb{E}[S|X=2] = 0.2 \qquad \mathbb{P}(X=2) = 0.5$$
$$\mathbb{E}[S|X=3] = 0 \qquad \mathbb{P}(X=3) = 0.2$$

Then

$$\begin{aligned}\mathbb{E}[S] &= \mathbb{E}[\mathbb{E}[S|X]] \\ &= \mathbb{P}(X=1)\mathbb{E}[S|X=1]+ \\ &\quad \mathbb{P}(X=2)\mathbb{E}[S|X=2]+ \\ &\quad \mathbb{P}(X=3)\mathbb{E}[S|X=3] \\ &= (0.3)(1)+(0.2)(0.5)+(0)(0.2) = \boxed{0.4000}.\end{aligned}$$

193. Food for Thought

The weather for a region is unusually rainy with probability 40%, in which case the chance of a farmer obtaining a full crop is 60%. It is normal precipitation with probability 40%, in which case there is a 90% chance of obtaining a full crop. Finally, there is a 20% chance of lower precipitation, which gives a 40% chance of a full crop.

a. Draw a probability tree for the probability of getting a full crop.
b. Calculate the overall probability of getting a full crop.

Solution
a. Let A denote the event that it rains. Then the probability tree is

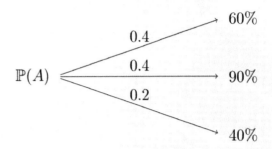

b. This is

$$\mathbb{P}(A) = (0.4)(60\%) + (0.4)(90\%) + (0.2)(40\%)$$
$$= \boxed{68\%}.$$

195. Random numbers of dice

Let $X_1, X_2, \ldots$ be iid $\text{Unif}(\{1,2,3\})$
a. What is $\mathbb{E}[X_i]$?
b. What is $\mathbb{E}[X_1+X_2+X_3+X_4]$?
c. If $G \sim \text{Geo}(0.3)$, what is

$$\mathbb{E}\left[\sum_{i=1}^{G} X_i\right].$$

Solution
a. This is $(1+3)/2 = \boxed{2}$.
b. Each of the four X_i have the same mean, and so summing them gives a mean of $4 \cdot 2 = \boxed{8}$.
c. Here we use the FTP to give:

$$\mathbb{E}\left(\sum_{i=1}^{G} X_i\right) = \mathbb{E}\left(\mathbb{E}\left(\sum_{i=1}^{G} X_i | G\right)\right)$$
$$= \mathbb{E}[2G] = 2/0.3 = 20/3 \approx \boxed{6.666}.$$

197. THE BAYESIAN, PART I

A Bayesian statistician models a parameter θ as having an Exp(2) distribution. The data X given θ is uniform over $[0, \theta]$.
a. What is $\mathbb{E}[X|\theta]$?
b. What is $\mathbb{E}[X]$?

SOLUTION
a. This will be
$$\frac{0+\theta}{2},$$
which is $\boxed{theta/2}$.
b. To undo the conditioning, we take the expectation again. Then we use the fact that the mean of an Exp(2) distribution is $1/2$ to get
$$\mathbb{E}[X] = (1/2)/2 = \boxed{0.2500}.$$

199. USING THE FTP

The FTP applies to multiple conditions as well. So for instance $\mathbb{E}[X] = \mathbb{E}[\mathbb{E}[X|Y, Z]]$.
Suppose $Z \sim \text{Unif}([1, 2])$, $[Y|Z] \sim \text{Exp}(Z)$ and $[X|Y, Z] \sim \text{Unif}(Z, Z+Y)$. Find $\mathbb{E}[X]$.

SOLUTION
Since $[X|Y, Z] \sim \text{Unif}(Z, Z+Y)$, we have
$$\mathbb{E}[X|Y, Z] = (Z + Y + Z)/2 = Z + Y/2.$$
Now we take the mean again to give
$$\mathbb{E}[X] = \mathbb{E}[Z] + \mathbb{E}[Y]/2.$$

So to solve the problem, we need $\mathbb{E}[Z]$ and $\mathbb{E}[Y]$. Since $Z \sim \text{Unif}([1, 2])$, $\mathbb{E}[Z] = (1+2)/2 = 1.5$. Using the FTP for Y:
$$\mathbb{E}[Y] = \mathbb{E}[\mathbb{E}[Y|Z]] = \mathbb{E}[1/Z],$$
which can be found with an integral
$$\mathbb{E}[1/Z] = \int_{\mathbb{R}} (1/z) \mathbb{I}(z \in [1, 2]) \, dz$$
$$= \int_1^2 1/z \, dz = \ln(2).$$

Putting it together,
$$\mathbb{E}[X] = 1.5 + \ln(2)/2 = \boxed{1.846\ldots}.$$

JOINT DENSITIES OF RANDOM VARIABLES

201. JOINT DENSITIES

Suppose (X, Y) have joint density
$$f_{(X,Y)}(x, y) = (12/5)(x^2 + xy)i(x, y)$$
where
$$i(x, y) = \mathbb{I}((x, y) \in [0, 1] \times [0, 1]).$$
a. Find $\mathbb{P}(X \leq 0.4, Y \leq 0.3)$.
b. Find the density of X, f_X.
c. Are X and Y independent?

SOLUTION
a. Using Tonelli's theorem,
$$p = \mathbb{P}(X \leq 0.4, Y \leq 0.3)$$
$$= (12/7) \int_{x=-\infty}^{0.4} \int_{y=-\infty}^{0.3} (x^2 + xy)i(x, y) \, dy \, dx$$
$$= (12/7) \int_{x=0}^{0.4} \int_{y=0}^{0.3} (x^2 + xy) \, dy \, dx$$
$$= (12/7) \int_{x=0}^{0.4} (x^2 y + xy^2/2)\big|_0^{0.3} \, dx$$
$$= (12/7) \int_{x=0}^{0.4} 0.3x^2 + 0.045x \, dx$$
$$= (12/7) \, 0.1x^3 + 0.0225x^2 \big|_0^{0.4},$$
which comes out to be $\boxed{1.714\%}$.
b. To find the density of X, integrate out the variable associated with Y:
$$f_X(x) = \int_{y=-\infty}^{\infty} f_{(X,Y)}(x, y) \, dy$$
$$= (12/7) \int_{y=-\infty}^{\infty} (x^2 + xy)i(x, y) \, dy$$
$$= (12/7)\mathbb{I}(x \in [0, 1]) \int_{y=0}^{1} x^2 + xy \, dy$$
$$= (12/7)\mathbb{I}(x \in [0, 1])x^2 y + xy^2/2\big|_0^1$$
$$= \boxed{(12/7)(x^2 + x/2)\mathbb{I}(x \in [0, 1])}.$$

c. To check independence, find the density Y by integrating out the variable associated with X.

$$f_Y(y) = \int_{x=-\infty}^{\infty} f_{(X,Y)}(x,y)\,dx$$
$$= (12/7)\int_{x=-\infty}^{\infty} (x^2 + xy)\mathbb{I}((x,y) \in [0,1]\times[0,1])\,dx$$
$$= (12/7)\mathbb{I}(y \in [0,1])\int_{x=0}^{1} x^2 + xy\,dx$$
$$= (12/7)\mathbb{I}(y \in [0,1])x^3/3 + x^2 y/2\big|_0^1$$
$$= (12/7)(1/3 + y/2)\mathbb{I}(y \in [0,1]).$$

Integrating then gives

$$\mathbb{P}(X \leq 1/2) = (12/7)\int_{x=0}^{1/2} x^2 + x/2\,dx$$
$$= 5/28$$

$$\mathbb{P}(Y \leq 1/2) = (12/7)\int_{y=0}^{1/2} 1/3 + y/2\,dy$$
$$= 11/28,$$

and

$$\mathbb{P}(X \leq 1/2, Y \leq 1/2) = \int_{x=0}^{1/2}\int_{y=0}^{1/2} (x^2 + xy)\,dy\,dx$$
$$= 1/16 = 0.0625,$$

while
$$\frac{5}{28} \cdot 1128 = 0.07015\ldots.$$

This proves the random variables are $\boxed{\text{not independent.}}$

203. INDEPENDENCE

Let X with density $f_X(s) = \exp(-2s)\mathbb{I}(s \geq 0)$ and Y with density $f_Y(r) = 2r\mathbb{I}(r \in [0,1])$ be independent random variables. What is the joint density

$$f_{(X,Y)}(s,r)?$$

SOLUTION

Because the random variables are independent, the joint density is the product of the individual densities.

$$\boxed{f_{(X,Y)}(s,r) = 2r\exp(-2s)\mathbb{I}(s \geq 0, r \in [0,1])}.$$

205. MULTIPLE INTEGRALS

Suppose the joint density of X_1 and X_2 is over the region $A = [0,1] \times [0,1]$.

$$f_{(X_1,X_2)}(x_1,x_2) = \frac{3}{8} \cdot \frac{x_1+x_2}{\sqrt{|x_1-x_2|}}\mathbb{I}((x_1,x_2) \in A).$$

Find $\mathbb{P}(X_1 > X_2 + 0.1)$ by setting up the integral and then using a numerical solver.

SOLUTION

The inequality $x_1 > x_2 + 0.1$ is the same as

$$x_2 < x_1 + 0.1.$$

Together with the inequalities $0 \leq x_1 \leq 1$ and $0 \leq x_2 \leq 1$, this gives the region:

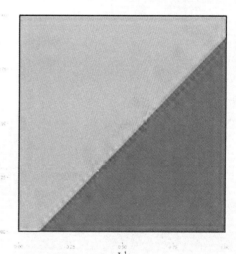

Let B be the region inside the triangle with vertices $(0.1, 0), (1, 0.9), (1, 0)$. Then

$$\mathbb{P}(X_1 > X_2 + 0.1) = \mathbb{P}((X_1,X_2) \in B).$$

Hence

$$\mathbb{P}(X_1 > X_2 + 0.1) = \mathbb{P}((X_1,X_2) \in B)$$
$$= \int_{(x_1,x_2)\in B} f_{(X_1,X_2)}(x_1,x_2)\,d\mathbb{R}^2$$
$$= \int_{(x_1,x_2)\in B} \frac{3}{8} \cdot \frac{x_1+x_2}{\sqrt{|x_1-x_2|}}\,d\mathbb{R}^2$$
$$= \int_{(x_1,x_2)\in B} \frac{3}{8} \cdot \frac{x_1+x_2}{\sqrt{x_1-x_2}}\,d\mathbb{R}^2,$$

where the last line follows since for $(x_1,x_2) \in B$, $|x_1-x_2| = x_1-x_2$.

CHAPTER 30: ENCOUNTERS RESOLVED

The integrand is nonnegative, so by Tonelli's Theorem:

$$\mathbb{P}(X_1 > X_2 + 0.1) = \int_{x_1=0}^{1} \int_{x_2=0}^{x_1-0.1} \frac{3}{8} \cdot \frac{x_1 + x_2}{\sqrt{x_1 - x_2}} \, dx_2 \, dx_1.$$

The integral is set up, now use a numerical solver such as Wolfram Alpha (www.wolframalpha.com)

```
integrate (3/8)*(x_1 + x_2)/
sqrt(x_1 - x_2) for x_1 from
0 to 1 and x_2 from 0 to x_1 - 0.1
```

This returns an answer of $\boxed{0.2707\ldots}$.

207. DISCRETE JOINT DENSITIES

Suppose (X, Y) are uniform over the four points $(-1, 1), (-1, -1), (0, 0), (1, 2)$.
a. What is the density of X?
b. What is the density of Y?
c. Are X and Y independent?

SOLUTION

a. To find the density of X, integrate out Y:

$$f_X(x) = \sum_{y \in \{-1,0,1,2\}} f_{(X,Y)}(x, y)$$

For $x = -1$, there are two y values, 1 and -1. Hence

$$f_X(-1) = (1/4) + (1/4) = (1/2)$$

For $x = 0$ and $x = 1$, there is exactly one y value that goes along with them, so

$$f_X(0) = f_X(1) = 1/4.$$

This makes the density $f_X(x)$ equal to

$$\boxed{\frac{1}{2}\mathbb{I}(x = -1) + \frac{1}{4}\mathbb{I}(x = 0) + \frac{1}{4}\mathbb{I}(x = 1)}.$$

b. To find the density of Y, integrate out the X:

$$f_Y(y) = \sum_{x \in \{-1,0,1,2\}} f_{(X,Y)}(x, y)$$

For each y value in $-1, 0, 1, 2$, there is precisely one x value with density $1/4$. Hence

$$\boxed{f_Y(y) = \frac{1}{4}\mathbb{I}(y \in \{-1, 0, 1, 2\})}.$$

That makes the distribution of Y equal to $\text{Unif}(\{-1, 0, 1, 2\})$.

c. Note

$$f_X(-1)f_Y(-1) = (1/2)(1/4) = 1/8,$$

but

$$f_{(X,Y)}(-1, -1) = 1/4.$$

Since these are different, X and Y are $\boxed{\text{not independent}}$.

209. FACTORING CONTINUOUS DENSITIES

Suppose $f_{(X,Y)}(x, y) = \mathbb{I}(x \in [0, 2])\exp(-2y)\mathbb{I}(y \geq 0)$. Show that X and Y are independent.

SOLUTION

Note that the joint density factors into

$$f_{(X,Y)}(x, y) = [\mathbb{I}(x \in [0, 2])][\exp(-2y)\mathbb{I}(y \geq 0)]$$

which is the product of a piece that only depends on x and a piece that only depends on y. Hence the random variables are independent.

211. DEPENDENT JOINT CONTINUOUS UNIFORMS

Suppose $X = (X_1, X_2)$ is uniform over the upper half of the unit circle given by

$$A = \{(x_1, x_2) : x_2 \geq 0, x_1^2 + x_2^2 \leq 1\}.$$

Since $\text{Leb}(A) = \pi/2$, this has density

$$f_{(X_1, X_2)}(x_1, x_2) = \frac{2}{\pi}\mathbb{I}(\{(x_1, x_2) \in A).$$

Show that X_1 and X_2 are not independent.

SOLUTION

If $X_2 \geq 1/2$, then $|X_1| \in [-\sqrt{3}/2, \sqrt{3}/2]$ with probability 1, and so

$$\mathbb{P}(|X_1| \leq \sqrt{3}/2, X_2 \geq 1/2) = \mathbb{P}(X_2 \geq 1/2).$$

On the other hand, $\mathbb{P}(|X_1| > \sqrt{3}/2) > 0$ since the area of this region is nonzero. Hence

$$\mathbb{P}(|X_1| \leq \sqrt{3}/2, X_2 \geq 1/2) \neq$$
$$\mathbb{P}(|X_1| \leq \sqrt{3}/2)\mathbb{P}(X_2 \geq 1/2).$$

Since the probability of the logical AND does not equal the product of the probabilities, the random variables are not independent.

RANDOM VARIABLES AS VECTORS

213. RULES OF INNER PRODUCTS

Say $\operatorname{cov}(X, Y) = 4.2$. Find $\operatorname{cov}(3X, -2Y)$.

SOLUTION
Factoring constants out of the covariance gives

$$\operatorname{cov}(3X, -2Y) = (3)(-2)\operatorname{cov}(X, Y) = -6 \cdot 4.2,$$

which is -25.20.

215. RULES OF INNER PRODUCT NORMS

Suppose that $\operatorname{sd}(X) = 1.8$.
a. What is the variance of X?
b. What is $\operatorname{sd}(3X)$?
c. What is $\operatorname{sd}(-3X)$?

SOLUTION
a. This is the standard deviation squared, which is 3.240.
b. Here $\operatorname{sd}(3X) = 3\operatorname{sd}(X)$, or 5.400.

c. Here $\operatorname{sd}(-3X) = -3\operatorname{sd}(X)$, or 5.400.

217. STANDARD DEVIATION OF DISCRETE RANDOM VARIABLES

Say X is discrete with density $f_X(-1) = 0.6$, $f_X(0) = 0.3$, $f_X(1) = 0.1$.
a. Find $\mathbb{E}[X]$.
b. Find $\operatorname{sd}(X)$.

SOLUTION
a. This is
$$\mathbb{E}[X] = (-1)(0.6) + (0)(0.3) + (1)(0.1),$$
so -0.5000.
b. For this, it is necessary to find $\mathbb{E}[X^2]$:
$$\mathbb{E}[X^2] = (-1)^2(0.6) + (0)^2(0.3) + (1)^2(0.1) = 0.7,$$
and so $\operatorname{sd}(X) = \sqrt{\mathbb{E}[X^2] - \mathbb{E}[X]^2}$, which is about 0.6708.

219. COVARIANCE OF A JOINT UNIFORM

Let (X, Y) be uniform over the triangle with vertices $(0,0)$, $(0,1)$, and $(1,0)$. Find $\operatorname{cov}(X, Y)$.

SOLUTION
The triangle with vertices $(0,0)$, $(0,1)$, and $(1,0)$ can be defined using the inequalities corresponding to the three sides:

$$T = \{(x, y) : x \geq 0, y \geq 0, x + y \leq 1\}.$$

This triangle has area $1/2$.
Hence

$$\mathbb{E}[XY] = \int_{\mathbb{R}^2} xy \frac{\mathbb{I}(x \geq 0, y \geq 0, x + y \leq 1)}{1/2} \, d\mathbb{R}^2$$

$$= 2 \int_{x=0}^{1} \int_{y=0}^{1-x} xy \, dy \, dx$$

$$= 2 \int_{x=0}^{1} xy^2/2 \Big|_0^{1-x} dx$$

$$= \int_{x=0}^{1} x(1-x)^2 \, dx$$

$$= \frac{\Gamma(2)\Gamma(3)}{\Gamma(2+3)} = \frac{1! \, 2!}{4!} = \frac{1}{12}.$$

CHAPTER 30: ENCOUNTERS RESOLVED

Similarly,

$$\mathbb{E}[X] = \int_{\mathbb{R}^2} x \frac{\mathbb{I}(x \geq 0, y \geq 0, x+y \leq 1)}{1/2} \, d\mathbb{R}^2$$

$$= \int_{x=0}^{1} \int_{y=0}^{1-x} 2xy \, dy \, dx$$

$$= \int_{x=0}^{1} 2x(1-x) \, dx$$

$$= 2\frac{\Gamma(2)\Gamma(2)}{\Gamma(2+2)} = 2\frac{1!1!}{3!} = \frac{1}{3},$$

and

$$\mathbb{E}[Y] = \int_{\mathbb{R}^2} y \frac{\mathbb{I}(x \geq 0, y \geq 0, x+y \leq 1)}{1/2} \, d\mathbb{R}^2$$

$$= \frac{1}{3},$$

as well.

Hence $\operatorname{cov}(X, Y) = 1/12 - (1/3)(1/3) = (3-4)/36 = -1/9$, or about $\boxed{-0.02777}$.

Solution
a. Integrate to get

$$\mathbb{E}[XY] = \int_{\mathbb{R}} (xy)2\mathbb{I}((x,y) \in A) \, dA$$

$$= \int_{x=0}^{1} \int_{y=x}^{1} 2xy \, dy \, dx$$

$$= \int_{x=0}^{1} xy^2\big|_{x}^{1} \, dx$$

$$= \int_{x=0}^{1} x - x^3 \, dx$$

$$= x^2/2 - x^4/4 \big|_0^1$$

$$= 1/4$$

so the answer is $\boxed{0.2500}$.

b. Integrate to get

$$\mathbb{E}[X] = \int_{\mathbb{R}} (x)2\mathbb{I}((x,y) \in A) \, dA$$

$$= \int_{x=0}^{1} \int_{y=x}^{1} 2x \, dy \, dx$$

$$= \int_{x=0}^{1} 2x(1-x) \, dx$$

$$= \int_{x=0}^{1} 2x - 2x^2 \, dx$$

$$= x^2 - 2x^3/3 \big|_0^1$$

$$= 1/3,$$

making the result $\boxed{0.3333\ldots}$.

c. Integrate to get

$$\mathbb{E}[Y] = \int_{\mathbb{R}} y \cdot 2\mathbb{I}((x,y) \in A) \, dA$$

$$= \int_{x=0}^{1} \int_{y=x}^{1} 2y \, dy \, dx$$

$$= \int_{x=0}^{1} y^2\big|_x^1 \, dx$$

$$= \int_{x=0}^{1} 1 - x^2 \, dx$$

$$= x - x^3/3 \big|_0^1$$

$$= 2/3,$$

givig $\boxed{0.6666\ldots}$ as the answer.

d. Putting together the pieces,

$$\operatorname{cov}(X, Y) = \mathbb{E}[XY] - \mathbb{E}[X]\mathbb{E}[Y]$$

$$= 1/4 - (1/3)(2/3)$$

$$= 9/36 - 8/36 = 1/36 = \boxed{0.02777\ldots}.$$

d. Find $\operatorname{cov}(S, T)$.

Correlation

223. Affine transforms

Suppose X and Y have correlation 0.4276. What is the correlation between $2X+4$ and $5Y+3$?

Solution
Note that $\operatorname{sd}(2X+4) = 2\operatorname{sd}(X)$, $\operatorname{sd}(5Y+3) = 5\operatorname{sd}(Y)$, and $\operatorname{cov}(2X+4, 5Y+3) = (2)(5)\operatorname{cov}(X,Y)$. Hence

$$\operatorname{cor}(2X+4, 5Y+3) = \frac{(2)(5)\operatorname{cov}(X,Y)}{2\operatorname{sd}(X)5\operatorname{sd}(Y)} = \operatorname{cor}(X,Y),$$

so the correction of the scaled and shifted random variables remains $\boxed{0.4276}$.

225. Correlation of discrete uniforms

For $(X, Y) \sim \mathsf{Unif}(\{(0,0), (2,0), (2,1)\})$, find the correlation between X and Y.

SOLUTION

Because there are three points in the set, each has $1/3$ chance of occurring. Hence

$$\mathbb{E}[X] = \frac{1}{3}(0) + \frac{1}{3}(2) + \frac{1}{3}(2) = \frac{4}{3}$$

$$\mathbb{E}[X^2] = \frac{1}{3}(0)^2 + \frac{1}{3}(2)^2 + \frac{1}{3}(2)^2 = \frac{8}{3}$$

$$\mathbb{E}[Y] = \frac{1}{3}(0) + \frac{1}{3}(0) + \frac{1}{3}(1) = \frac{1}{3}$$

$$\mathbb{E}[Y^2] = \frac{1}{3}(0)^2 + \frac{1}{3}(0)^2 + \frac{1}{3}(1)^2 = \frac{1}{3}$$

$$\mathbb{E}[XY] = \frac{1}{3}(0)(0) + \frac{1}{3}(2)(0) + \frac{1}{3}(2)(1) = \frac{2}{3}.$$

Hence

$$\operatorname{cor}(X, Y) = \frac{\operatorname{cov}(X, Y)}{\operatorname{sd}(X) \operatorname{sd}(Y)}$$

$$= \frac{\mathbb{E}[XY] - \mathbb{E}[X]\mathbb{E}[Y]}{\sqrt{\mathbb{E}[X^2] - \mathbb{E}[X]^2}\sqrt{\mathbb{E}[Y^2] - \mathbb{E}[Y]^2}}$$

$$= \frac{(2/3) - (4/3)(1/3)}{\sqrt{8/3 - (4/3)^2}\sqrt{1/3 - (1/3)^2}},$$

which is $\boxed{0.5000}$.

227. Understanding Functions

Suppose $U \sim \operatorname{Unif}([0, 1])$. Find $\operatorname{cor}(U, U^2)$.

SOLUTION

By symmetry:

$$\mathbb{E}[U] = \frac{0+1}{2} = \frac{1}{2},$$

by integrating

$$\mathbb{E}[U^2] = \int_0^1 u^2 \, du = \frac{1}{3}u^3 \Big|_0^1 = \frac{1}{3}.$$

Again by integrating

$$\mathbb{E}[U \cdot U^2] = \int_0^1 u^3 \, du = \frac{1}{4}.$$

Similarly,

$$\mathbb{E}[(U^2)^2] = \mathbb{E}[U^4] = \frac{1}{5}.$$

Hence the correlation is

$$\operatorname{cor}(U, U^2) = \frac{(1/4) - (1/2)(1/3)}{\sqrt{(1/3) - (1/2)^2}\sqrt{(1/5) - (1/3)^2}}$$

which is about $\boxed{0.9682}$.

229. The Pythagorean Theorem

Suppose S, T, R are independent random variables with variances of $1.1, 2.8, 0.6$ respectively.
a. What is $\operatorname{var}(S + T + R)$?
b. What is $\operatorname{var}(S - 2T)$?

SOLUTION

a. Because they are independent, this is $1.1 + 2.8 + 0.6$, or $\boxed{4.500}$.
b. If S and T are independent, S and $-2T$ are as well. Hence this is

$$\operatorname{var}(S) + (-2)^2 \operatorname{var}(T),$$

which is $\boxed{12.30}$.

231. Sample Averages

Suppose X has standard deviation 3.2. What is the standard deviation of $(X_1 + \cdots + X_{100})/100$ if the X_i are iid X?

SOLUTION

This is $\operatorname{sd}(X)/\sqrt{100}$, or $\boxed{0.3200}$.

Suppose Y has standard deviation 0.4. How big does n need to be for the standard deviation of $(Y_1 + \cdots + Y_n)/n$ to be at most 0.01?

Adding Random Variables

233. Adding random variables

Suppose $(X, Y) \sim \operatorname{Unif}(\{(1, 2), (1, 3), (2, 2)\})$. What is the density of $X + Y$?

Solution

There is one point where $X+Y=3$, and two points where $X+Y=4$. Therefore $\mathbb{P}(X+Y=3) = 1/3$, and $\mathbb{P}(X+Y=4) = 2/3$. So the density of $X+Y$ is

$$\boxed{f_{X+Y}(s) = \frac{1}{3}\mathbb{I}(s=3) + \frac{2}{3}\mathbb{I}(s=4).}$$

235. Adding Independent Dice

Suppose $A \sim$ d4 and $B \sim$ d4 are independent. Use the convolution of the densities of A and B to find the density of $A+B$.

Solution
The convolution of the densities of A and B gives

$$f_{A+B}(s) = \int_a f_A(a) f_B(s-a) \, d\#$$
$$= \sum_a f_A(a) f_B(s-a)$$
$$= \sum_{a \in \mathbb{Z}} \frac{1}{4}\mathbb{I}(1 \leq a \leq 4) \frac{1}{4}\mathbb{I}(1 \leq s-a \leq 4)$$
$$= \sum_{a \in \mathbb{Z}} \frac{1}{4}\mathbb{I}(1 \leq a \leq 4) \frac{1}{4}\mathbb{I}(s-4 \leq a \leq s-1)$$
$$= \frac{1}{16} \sum_a \mathbb{I}(\max(1, s-4) \leq a \leq \min(4, s-1)).$$

This is 0 if $s-1 < 1$ (so $s < 2$) or $s-4 > 4$ (so $s > 8$). If $s \in \{2, \ldots, 8\}$, then this is

$$\min(4, s-1) - \max(1, s-4) + 1,$$

which if $s-4 > 1$ is $4-(s-4)+1 = 9-s$, and if $s-4 < 1$ it is $s-1-1+1 = s-1$.

Putting this all together gives

$$\boxed{f_{A+B}(s) = \frac{s-1}{16}\mathbb{I}(s \in \{2,3,4\}) + \frac{9-s}{16}\mathbb{I}(s \in \{5,6,7,8\}).}$$

237. Adding Continuous Random Variables

Suppose $A \sim \text{Unif}([0,1])$ and $B \sim \text{Exp}(1)$. What is the density of $A+B$?

Solution
Find the convolution of the densities of A and B:

$$f_{A+B}(s) = \int_a f_A(a) f_B(s-a) \, ds$$
$$= \int_a \mathbb{I}(a \in [0,1]) \exp(-(s-a))\mathbb{I}(s-a \geq 0)$$
$$= \int_a \mathbb{I}(0 \leq a \leq 1)\mathbb{I}(a \leq s) \exp(-(s-a)) \, da.$$

If $s < 0$, the indicators will always be empty. If $s \in [0,1]$, then a runs from 0 up to s. If $s > 1$, then a runs from 0 up to 1. So for $s \in [0,1]$,

$$f_{A+B}(s) = \int_{a=0}^{s} \exp(-(s-a)) \, da$$
$$= \exp(-(s-a))\Big|_0^s$$
$$= 1 - \exp(-s).$$

For $s > 1$,

$$f_{A+B}(s) = \int_{a=0}^{1} \exp(-(s-a)) \, da$$
$$= \exp(-(s-a))\Big|_0^1$$
$$= \exp(-s+1) - \exp(-s)$$
$$= \exp(-s)[e-1].$$

Hence the complete density is

$$\boxed{\begin{aligned}f_{A+B}(s) = &[1-\exp(-s)]\mathbb{I}(s \in [0,1]) + \\ &\exp(-s)[e-1]\mathbb{I}(s > 1).\end{aligned}}$$

239. Generating Function of a Poisson

Suppose $X \sim \text{Pois}(\mu)$. Prove that the generating function of X is $\exp(-\mu(1-x))$.

Solution

Recall that for $X \sim \text{Pois}(\mu)$,

$$\mathbb{P}(X = i) = \frac{\exp(-\mu)\mu^i}{i!}\mathbb{I}(i \in \{0, 1, 2, \ldots\}).$$

Hence

$$\begin{aligned}
\text{gf}_X(x) &= \mathbb{E}[x^X] \\
&= \sum_{i=0}^{\infty} x^i \exp(-\mu)\mu^i/i! \\
&= \sum_{i=0}^{\infty} \exp(-\mu)[x\mu]^i/i! \\
&= \exp(-\mu)\exp(x\mu), \\
&= \exp(-\mu(1-x)),
\end{aligned}$$

which completes the proof. □

241. Wolfram Alpha to the Rescue

Suppose $X_1, \ldots, X_{10}$ are iid d4. Then using Wolfram Alpha to perform the polynomial multiplications, find the probability that $X_1 + \cdots + X_{10} = 23$.

Solution
Typing

```
expand ((1/4)*(x+x^2+x^3+x^4))^10
```

into Wolfram Alpha gives a x^{23} term of $(50055/524288)x^{23}$, hence the probability is about $\boxed{0.09547}$.

The Central Limit Theorem

243. CLT for Exponentials

Suppose $T_1, T_2, \ldots, T_{18}$ are iid $\text{Exp}(3.4)$. Estimate $\mathbb{P}(T_1 + \cdots + T_{18} \geq 6)$ using the Central Limit Theorem.

Solution
First calculate the mean and variance of the T_i: $\mathbb{E}[T_i] = 1/3.4$ and $\text{var}(T_i) = 1/3.4^2$. Then

$$\begin{aligned}
p &= \mathbb{P}(T_1 + \cdots + T_{18} \geq 6) \\
&= \mathbb{P}(T_1 + \cdots + T_{18} - 18/3.4 \geq 6 - 18/3.4) \\
&= \mathbb{P}(T_1 + \cdots + T_{18} - 18/3.4 \geq 6 - 18/3.4) \\
&= \mathbb{P}\left(\frac{T_1 + \cdots + T_{18} - 18/3.4}{\sqrt{18/3.4^2}} \geq \frac{6 - 18/3.4}{\sqrt{18/3.4^2}}\right) \\
&\approx \mathbb{P}(Z \geq 0.5656854),
\end{aligned}$$

which is $\boxed{0.2858}$.

245. CLT for Geometrics

Suppose $T_1, T_2, \ldots, T_{18}$ are iid $\text{Geo}(0.2)$. Estimate $\mathbb{P}(T_1 + \cdots + T_{18} \geq 100)$ using the Central Limit Theorem and the half-integer correction.

Solution
The mean of the T_i is $1/0.2 = 5$, and the variance is $(1-0.2)/(0.2)^2 = 20$. So

$$\begin{aligned}
p &= \mathbb{P}(T_1 + \cdots + T_{18} \geq 100) \\
&= \mathbb{P}(T_1 + \cdots + T_{18} \geq 99.5) \\
&= \mathbb{P}(T_1 + \cdots + T_{18} - 18 \cdot 5 \geq 99.5 - 18 \cdot 5) \\
&= \mathbb{P}\left(\frac{T_1 + \cdots + T_{18} - 18 \cdot 5}{\sqrt{18 \cdot 20}} \geq \frac{99.5 - 18 \cdot 5}{\sqrt{18 \cdot 20}}\right) \\
&\approx \mathbb{P}(Z \geq 0.500694),
\end{aligned}$$

which is about $\boxed{0.3082}$.

247. Another Exponential CLT

Suppose $T_1, \ldots, T_{10}$ are iid $\text{Exp}(1.4)$. Estimate $\mathbb{P}(T_1 + \cdots + T_{10} \geq 9)$.

Solution
First standardize the sum *inside* the probability. For $S = T_1 + \cdots + T_{10}$,

$$\begin{aligned}
\mathbb{P}(S \geq 9) &= \mathbb{P}(S - 10 \cdot (1/1.4) \geq 9 - 10/1.4) \\
&= \mathbb{P}\left(\frac{S - 10 \cdot (1/1.4)}{\sqrt{10}(1/1.4)} \geq \frac{9 - 10/1.4}{\sqrt{10}(1/1.4)}\right) \\
&\approx \mathbb{P}(Z \geq 0.8221922),
\end{aligned}$$

which is $\boxed{0.2054}$.

NORMAL RANDOM VARIABLES

253. SHIFTING AND SCALING NORMALS

Suppose $W \sim \mathsf{N}(34, 20)$. Then write $\mathbb{P}(W \leq 28)$ in terms of cdf_Z where $Z \sim \mathsf{N}(0, 1)$.

SOLUTION
Recall $W \sim 34 + \sqrt{20}Z$, so

$$\mathbb{P}(W \leq 28) = \mathbb{P}(34 + \sqrt{20}Z \leq 28)$$
$$= \mathbb{P}(\sqrt{20}Z \leq -6)$$
$$= \mathbb{P}(Z \leq -6/\sqrt{20}),$$

hence $\mathbb{P}(W \leq 28)$ equals $\boxed{\mathrm{cdf}_Z(-1.341)}$.

255. ADDING INDEPENDENT NORMALS

If $Z_1 \sim \mathsf{N}(2.4, 5.2)$ and $Z_2 \sim \mathsf{N}(-1.2, 5.2)$ are independent, what is the distribution of $Z_1 + Z_2$?

SOLUTION
The sum of normals is itself a normally distributed random variable. Sum the means and variances to get the new mean and variance. So $\boxed{Z_1 + Z_2 \sim \mathsf{N}(1.2, 10.4)}$.

257. MULTIVARIATE SCALING AND SHIFTING

Suppose that (Z_1, Z_2) are iid standard normal random variables, and

$$W_1 = 3Z_1 - Z_2 + 3$$
$$W_2 = -2Z_1 + Z_2 - 4.$$

What is the distribution of (W_1, W_2)?

SOLUTION
The family of distributions is Multinormal, now to find the parameters. The mean of (W_1, W_2) is $(3, -4)$, and the covariance matrix is

$$\begin{pmatrix} 3 & -1 \\ -2 & 1 \end{pmatrix} \begin{pmatrix} 3 & -2 \\ -1 & 1 \end{pmatrix} = \begin{pmatrix} 10 & -7 \\ -7 & 5 \end{pmatrix}$$

Hence

$$\boxed{\begin{pmatrix} W_1 \\ W_2 \end{pmatrix} \sim \mathsf{Multinormal}\left(\begin{pmatrix} 3 \\ -4 \end{pmatrix}, \begin{pmatrix} 10 & -7 \\ -7 & 5 \end{pmatrix}\right)}$$

BAYES RULE FOR DENSITIES

259. ARCHYTAS MEDICAL

Archytas Medical Group believes a new drug has p chance of working, where p is a random variable with density $f_p(a) \sim \mathsf{Beta}(2, 3)$. They test the drug on five animals, three of whom show that the drug works. What is the posterior distribution on p given this information?

SOLUTION
The prior density is

$$f_p(a) = C_1 a(1-a)^2 \mathbb{I}(a \in [0, 1]).$$

Let N denote the number of successes. Then the density of N gives $p = a$ at 3 is

$$f_{N|p=a}(3) = C_2 a^3 (1-a)^2.$$

The product of these two is proportional to the posterior:

$$f_{p|N=3}(a) = C_3 a^4 (1-a)^4 \mathbb{I}(a \in [0, 1]).$$

Hence $\boxed{[p \mid N = 3] \sim \mathsf{Beta}(5, 5)}$.

261. TOOTHPASTE TROUBLES

A factory produces 980 tubes of toothpaste in a day. The chance that any tube is defective is 0.001.

a. What is the chance that no tubes of toothpaste are defective?
b. What is the chance that at least two tubes of toothpaste are defective?

SOLUTION
Let D be the number of defective tubes. Then
$$D \sim \text{Bin}(980, 0.001).$$

a. The probability a binomial is 0 is
$$\mathbb{P}(D=0) = \binom{980}{0}(0.001)^0(0.999)^{980}.$$

Note that 980 choose 0 is 1 (there is only one way for all 980 trials to be a failure) and so the answer is about $\boxed{0.3751}$.

b. Using the complement rule, our target event $d = (D \geq 2)$ has probability
$$\begin{aligned}\mathbb{P}(d) &= 1 - \mathbb{P}(D \leq 1)\\&= 1 - \mathbb{P}(D=0) - \mathbb{P}(D=1)\\&= 1 - 0.999^{980} - \binom{980}{1}(0.001)^1(0.999)^{979}\\&= 1 - 0.999^{980} - 980(0.001)^1(0.999)^{979}\\&= 0.2568\ldots,\end{aligned}$$

so the answer is $\boxed{0.2568}$.

263. RONCO SURVEY GROUP

Ronco Survey Group knows that the chance of someone answering their phone and doing a survey is 4%. How many people do they have to call in order to make sure that the probability that they get at least 10 survey takers is at least 70

SOLUTION
Start with $n = 10/0.04 = 250$:

`1 - pbinom(9, 250, 0.04)`

`## [1] 0.544631`

Too small. Try $n = 300$.

`1 - pbinom(9, 300, 0.04)`

`## [1] 0.762957`

Too big! Continue to narrow in on the right number until

`1 - pbinom(9, 283, 0.04)`

`## [1] 0.6980266`

and

`1 - pbinom(9, 284, 0.04)`

`## [1] 0.7021327`

indicate that $\boxed{284}$ is the correct result.

265. DIMER MEDICINE

Dimer Medicine creates 3 types of drugs for a particular illness. The first is effective in 50% of patients, the second in 37%, and the third in 5%. Let A denote the event that the drug is effective and $X \in \{1, 2, 3\}$ the drug that is given to the patient.

a. If a patient is equally likely to receive any of the three drugs, find the probability that both drug 1 is administered and it is effective.
b. If a patient is equally likely to receive any of the three drugs, what is the probability that the drug is effective on their illness? Hint: the event A is the disjoint union of three pieces.

$$A = (A \cap \{X=1\}) \cup (A \cap \{X=2\}) \cup (A \cap \{X=3\}).$$

c. If the drug is effective for the patient, what is the probability that the drug was of the third type?

CHAPTER 30: ENCOUNTERS RESOLVED

Solution

a. Let X be the type of drug administered 1, 2, or 3, and let A denote the event that the drug is effective. Since each patient is equally likely to receive any of the drugs, $\mathbb{P}(X = 1) = 1/3$. Also $\mathbb{P}(A \mid X = 1) = 0.5$. Hence

$$\mathbb{P}(A, X = 1) = \mathbb{P}(X = 1)\mathbb{P}(A \mid X = 1) = (1/3)(0.5)$$

which is $\boxed{0.1666}$.

b. Per the hint, break the event into three disjoint pieces:

$$A = \{A, X = 1\} \cup \{A, X = 2\} \cup \{A, X = 3\}.$$

Then finding the probability of each piece as in part (a) gives

$$\mathbb{P}(A) = (0.5)/3 + (0.37)/3 + (0.05)/3$$
$$= 0.92/3$$

which is $\boxed{0.3066}$.

c. Bayes' Rule!

$$\mathbb{P}(X = 3 \mid A) = \mathbb{P}(A \mid X = 3) \cdot \frac{\mathbb{P}(X = 3)}{\mathbb{P}(A)}$$
$$= 0.05 \cdot \frac{1/3}{0.92/3} = \frac{0.05}{0.92}.$$

which is $\boxed{0.05434}$.

267. The Assembly Line

A consultant models the probability p that an item is defective on an assembly line as being uniform over the set $\{0, 0.01, 0.02, 0.03, 0.04, 0.05\}$. After testing 100 items that are believed to be independently defective or not defective, 4 are found to be defective. What is the distribution of p conditioned on this information?

Solution

Let N denote the number of defective items. Then

$$[N \mid p] \sim \text{Bin}(100, p).$$

Let $\alpha \in \{0, 0.01, \ldots, 0.05\}$. Letting r be the result that we are looking for, Bayes' Rule gives

$$r = \mathbb{P}(p = \alpha \mid N = 4)$$
$$= \mathbb{P}(N = 4 \mid p = \alpha)\frac{\mathbb{P}(p = \alpha)}{\mathbb{P}(N = 4)}$$
$$= C\alpha^4(1-\alpha)^{100-4}\mathbb{I}(\alpha \in \{0, 0.01, \ldots, 0.05\}).$$

Since probabilities must add to 1,

$$C \sum_{\alpha \in \{0, 0.01, \ldots, 0.05\}} \alpha^4(1-\alpha)^{96} = 1,$$

which gives $C = 6002306$.

Hence the full posterior distribution is:

$$\mathbb{P}(p = \alpha \mid N = 4) = 6002306\alpha^4(1-\alpha)^{96},$$

which leads to the following table.

$\mathbb{P}(p = 0.00 \mid N = 4) = 0$
$\mathbb{P}(p = 0.01 \mid N = 4) = 0.02287\ldots$
$\mathbb{P}(p = 0.02 \mid N = 4) = 0.1380\ldots$
$\mathbb{P}(p = 0.03 \mid N = 4) = 0.2611\ldots$
$\mathbb{P}(p = 0.04 \mid N = 4) = 0.3052\ldots$
$\mathbb{P}(p = 0.05 \mid N = 4) = 0.2726\ldots$

269. Rush Hour

In a particular county during rush hour, 80% of cars contain one occupant, 10% contain 2, 5% contain 3, and 5% contains 4 or more.

Any car containing two or more occupants has a 90% chance of using the carpool lane, and 1% of cars containing only one occupant cheat and use the carpool lane.

Suppose a car is in the carpool lane. Given this information, what is the probability that the car contains 1, 2, 3, or 4+ occupants?

SOLUTION

Let $N \in \{1, 2, 3, 4+\}$ be the number of people in the car. Then from the information in the problem

$$\mathbb{P}(N = 1) = 0.8$$
$$\mathbb{P}(N = 2) = 0.1$$
$$\mathbb{P}(N = 3) = 0.05$$
$$\mathbb{P}(N = 4+) = 0.05$$

Let HOV be the event that the car uses the carpool (the high-occupancy-vehicle or HOV) lane. Then

$$\mathbb{P}(N = 1, HOV) = (0.8)(0.01)$$
$$\mathbb{P}(N = 2, HOV) = (0.1)(0.9)$$
$$\mathbb{P}(N = 3, HOV) = (0.05)(0.9)$$
$$\mathbb{P}(N = 4+, HOV) = (0.05)(0.9)$$

Adding these numbers gives 0.188. Hence Bayes' Rule gives

$$\mathbb{P}(N = 1 \mid HOV) = (0.8)(0.01)(0.188)^{-1}$$
$$\mathbb{P}(N = 2 \mid HOV) = (0.1)(0.9)(0.188)^{-1}$$
$$\mathbb{P}(N = 3 \mid HOV) = (0.05)(0.9)(0.188)^{-1}$$
$$\mathbb{P}(N = 4+ \mid HOV) = (0.05)(0.9)(0.188)^{-1}$$

Putting this to 4 sig figs gives

$$\boxed{\begin{aligned}\mathbb{P}(N = 1 \mid HOV) &= 0.04255\\ \mathbb{P}(N = 2 \mid HOV) &= 0.4787\\ \mathbb{P}(N = 3 \mid HOV) &= 0.2393\\ \mathbb{P}(N = 4+ \mid HOV) &= 0.2393\end{aligned}}$$

THE MULTINOMIAL DISTRIBUTION

271. PROBABILITIES FOR MULTINOMIALS

If $(X_1, X_2, X_3, X_4) \sim$ Multinomial$(10, 0.3, 0.5, 0.1, 0.1)$, what is the chance that (X_1, X_2, X_3, X_4) equals $(3, 5, 1, 1)$?

SOLUTION

Using the density formula, this is

$$f_{(X_1,\ldots,X_4)}(3, 5, 1, 1) = \binom{10}{3, 5, 1, 1} 0.3^3 0.5^5 0.1^1 0.1^1,$$

which is about $\boxed{0.04252}$.

273. MEANS OF MULTINOMIALS

If $(X_1, X_2, X_3, X_4) \sim$ Multinomial$(10, 0.3, 0.5, 0.1, 0.1)$, what is $\mathbb{E}[(X_1, X_2, X_3, X_4)]$?

SOLUTION

Marginally, each X_i is binomial with parameters 10 and $(0.3, 0.5, 0.1, 0.1)$, so the mean of the vector is:

$$\boxed{\mathbb{E}[(X_1, X_2, X_3, X_4)] = (3, 5, 1, 1).}$$

275. MULTINOMIAL COVARIANCE

If $(X_1, X_2, X_3, X_4) \sim$ Multinomial$(10, 0.3, 0.5, 0.1, 0.1)$, what is the covariance matrix for (X_1, X_2, X_3, X_4)?

SOLUTION

Recall that $\text{var}(X_i) = np_i(1 - p_i)$ and $\text{cov}(X_i, X_j) = -np_ip_j$ for multinomials. Hence the covariance matrix for this problem is (factoring out the 10)

$$10 \begin{pmatrix} (0.3)(0.7) & -(0.3)(0.5) & -(0.3)(0.1) & -(0.3)(0.1) \\ -(0.5)(0.3) & (0.5)(0.5) & -(0.5)(0.1) & -(0.5)(.1) \\ -(0.1)(0.3) & -(0.1)(0.5) & (0.1)(0.9) & -(0.1)(0.1) \\ -(0.1)(0.3) & -(0.1)(0.5) & -(0.1)(0.1) & (0.1)(0.9) \end{pmatrix}$$

which evaluates to

$$\boxed{\begin{pmatrix} 2.1 & -1.5 & -0.3 & -0.3 \\ -1.5 & 2.5 & -0.5 & -0.5 \\ -0.3 & -0.5 & 0.9 & -0.1 \\ -0.3 & -0.5 & -0.1 & 0.9 \end{pmatrix}}.$$

TAIL INEQUALITIES

277. MARKOV'S INEQUALITY

Suppose $T \geq 0$ has $\mathbb{E}[T] = 2.3$. Upper bound $\mathbb{P}(T \geq 6)$.

SOLUTION
Since $T \geq 0$, $\mathtt{T} = T$. So by Markov's inequality,

$$\mathbb{P}(T \geq 6) \leq \mathbb{E}[T]/6 = 2.3/6,$$

which is about $\boxed{0.3833}$.

279. CHEBYSHEV'S INEQUALITY

Suppose R has mean 12.2 and standard deviation 4.3. Use Chebyshev's inequality to bound $\mathbb{P}(R \leq 4)$.

SOLUTION
First put the problem in a form for Chebyshev:

$$\begin{aligned}(R \leq 4) &= (R - 12.2 \leq -8.2) \\ &= (-(R - 12.2) \geq 8.2) \\ &\to (|R - 12.2| \geq 8.2)\end{aligned}$$

Applying Chebyshev then gives

$$\begin{aligned}\mathbb{P}(R \leq 4) &\leq \mathbb{P}(|R - 12.2| \geq 8.2) \\ &\leq \frac{\mathrm{var}(R)}{8.2^2} \\ &= \frac{4.3^2}{8.2^2},\end{aligned}$$

which is at most $\boxed{0.2750}$.

281. CHEBYSHEV VIA STANDARD DEVIATION

What is the largest chance possible that a random variable is at least 3 standard deviations away from its mean?

SOLUTION
This is at most $(1/3)^2$, or at most $\boxed{0.1112}$ to four significant digits.

283. SAMPLE AVERAGES AND CHEBYSHEV

Suppose $N_1, N_2, \ldots$ are iid N, where $\mathbb{E}[N] = 3.2$ and $\mathrm{sd}(N) = 6.2$. Let $S_n = (N_1 + \cdots + N_n)/n$. How large must n be in order for Chebyshev to upper bound $\mathbb{P}(S_n < 0)$ by 0.2?

SOLUTION
Using Chebyshev

$$\begin{aligned}\mathbb{P}(S_n \leq 0) &= \mathbb{P}\left(\frac{S_n - 3.2}{6.2} \leq \frac{0 - 3.2}{6.2}\right) \\ &\leq \mathbb{P}\left(\left|\frac{S_n - 3.2}{6.2}\right| \geq \frac{3.2}{6.2}\right) \\ &= \frac{1}{n(3.2/6.2)^2}.\end{aligned}$$

Setting $1/[n(3.2/6.2)^2] \leq 0.2$ gives $n \geq 18.76\ldots$, so $\boxed{19}$.

285. CONSTRUCTION WOES

A building is believed to require 0.8 years on average to complete, with a standard deviation of 0.5 years. Give the best bound (either Markov or Chebyshev) for the following.
a. The building takes at least a year to build.
b. The building takes at least two years to build.

SOLUTION
Let C be the construction time of the building. Then $\mathbb{E}[C] = 0.8$ and $\mathrm{sd}(C) = 0.5$.
a. Markov's inequality gives

$$\mathbb{P}(C \geq 1) \leq 0.8/1 = 80\%.$$

Chebyshev's inequality gives

$$\mathbb{P}(C \geq 1) = \mathbb{P}(C - 0.8 \geq 0.2) \leq \frac{0.5^2}{0.2^2},$$

so leaves 1 as the best bound. Hence the best bound from this information is $\boxed{80\%}$.

b. Markov's inequality gives

$$\mathbb{P}(C \geq 2) \leq 0.8/2 = 40\%.$$

Chebyshev's inequality gives

$$\mathbb{P}(C \geq 2) = \mathbb{P}(C - 0.8 \geq 1.2) \leq \frac{0.5^2}{1.2^2} = 0.1736111$$

making the best upper bound $\boxed{17.37\%}$.

CHERNOFF INEQUALITIES

287. CHERNOFF FOR A POISSON

For $X \sim \text{Pois}(21.3)$, use Chernoff's inequality to bound $\mathbb{P}(X \geq 30)$.

SOLUTION
Using the result from earlier

$$\mathbb{P}(X \geq 30) \leq \exp(30 - 21.3)/(30/21.3)^{30},$$

which is at most $\boxed{0.2071}$.

289. CHERNOFF FOR GAMMAS

Using $t = 0.47$ in Chernoff's bound, give an upper bound for $R \sim \text{Gamma}(13, 1.4)$ of $\mathbb{P}(R \geq 14)$.

SOLUTION
Let $T_1, \ldots, T_{13} \sim \text{Exp}(1.4)$. Then $T_1 + \cdots + T_{13} \sim R$. Also,

$$\text{mgf}_{T_1}(t) = \int_s \exp(ts)\lambda \exp(-\lambda s) \mathbb{I}(s \geq 0)\, ds$$
$$= \int_{s \geq 0} \lambda \exp(-s(\lambda - t))\, ds$$
$$= \frac{\lambda}{\lambda - t}$$

when $t < \lambda$. Hence for $R \sim \text{Gamma}(13, 1.4)$,

$$\text{mgf}_R(0.47) = \left(\frac{1.4}{1.4 - 0.47}\right)^{13}$$

and

$$\mathbb{P}(R \geq 14) \leq \frac{(1.4/(1.4 - 0.47))^{13}}{\exp(0.47 \cdot 14)},$$

which is at most $\boxed{0.2830}$.

291. CHERNOFF GIVEN THE MGF

Suppose that X has $\text{mgf}_X(0.2)/\exp(4(0.2)) \leq 0.6$, and $X_1, \ldots, X_{10}$ are iid X. What is a bound on

$$\mathbb{P}\left(\frac{X_1 + \cdots + X_{10}}{10} \geq 4\right)?$$

SOLUTION
This is 0.6^{10}, or $\boxed{0.006047}$.

293. CHERNOFF FOR UNIFORMS

For $U_1, \ldots, U_n$ iid $\text{Unif}([0, 1])$, consider

$$P((U_1 + \cdots + U_{10}) \geq 6).$$

a. Use Wolfram Alpha to find the value of t that minimizes the Chernoff bound for this probability.
b. Find the best Chernoff bound for this probability.

SOLUTION
The moment generating function of a standard uniform is

$$\text{mgf}_U(t) = \mathbb{E}[\exp(tU)]$$
$$= \int_u \exp(tu)\mathbb{I}(u \in [0, 1])\, du$$
$$= \int_{u=0}^{1} \exp(tu)\, du$$
$$= \begin{cases} 1 & t = 0 \\ \frac{e^t - 1}{t} & t \neq 0. \end{cases}$$

CHAPTER 30: ENCOUNTERS RESOLVED

For summing variables,
$$s = \mathbb{P}(U_1 + \cdots + U_{10} \geq 6)$$
$$= \mathbb{P}((U_1 + \cdots + U_n)/10 \geq 0.6)$$
$$\leq [\mathrm{mgf}_U(t) \exp(-0.6t)]^{10}.$$

a. To minimize the Chernoff bound, ignore the exponent of 10 and minimize
$$g(t) = \frac{e^t - 1}{t} \exp(-0.6t) = \frac{\exp(0.4t) - \exp(-0.6t)}{t}.$$

Then the minimum is at about $\boxed{1.229}$.

b. Evaluating g at this t and raising to the 10th power gives an upper bound (to four sig figs) of $\boxed{0.5448}$.

THE HYPERGEOMETRIC DISTRIBUTION

295. THOSE CRAZY EIGHTS

In a standard 52 deck of cards, there are four cards with rank 8 (the 8 of hearts, the 8 of spades, the 8 of diamonds, and the 8 of clubs).

If seven cards are dealt out uniformly at random from the deck, what is the chance that exactly two are rank 8?

SOLUTION
The number of cards of rank 8 will be hypergeometric
$$[N_7 \mid N_{52} = 4] \sim \mathsf{HyperGeo}(52, 7, 4)$$

Hence
$$\mathbb{P}(N_7 = 2 \mid N_{52} = 4) = \frac{\binom{7}{2}\binom{52-4}{4-2}}{\binom{52}{4}}$$
$$= \frac{7!45!}{2!5!2!43!} \cdot \frac{4!48!}{52!}$$
$$= \frac{7 \cdot 6 \cdot 45 \cdot 44 \cdot 4 \cdot 3}{2 \cdot 1 \cdot 52 \cdot 51 \cdot 50 \cdot 49}$$
$$= \boxed{0.07679\ldots}.$$

297. A BAG OF MARBLES

A bag contains five red and ten blue marbles. If four marbles are selected at random, what is the chance that exactly three are red?

SOLUTION
From the first expression, $k = 4$, $\ell = 5$, $j = 3$, $n = 5 + 10 = 15$. So
$$\mathbb{P}(H = 3) = \binom{4}{3}\frac{5^{\underline{3}} 10^{\underline{1}}}{15^{\underline{4}}} = \frac{50}{273},$$
or about $\boxed{0.07326}$.

299. A BOX OF SCREWS

A box of screws contains 40 type A and 60 type B screws. If a group of 30 screws is chosen uniformly at random from the box, then what is the chance that the last screw chosen is type A?

SOLUTION
Whether the first screw chosen or the last screw chosen, this will be 40/100, or $\boxed{40\%}$.

301. THE MISSION

Three out of eight members of the Ranger's Guild are Wood Elves. If five of the members are chosen for a secret mission uniformly at random, what is the chance that exactly two are Wood Elves?

SOLUTION
The number of Wood Elves chosen is
$$C \sim \mathsf{HyperGeo}(8, 3, 5),$$

Then the probability that the first two drawn are Wood Elves and the next three are not will be
$$\mathbb{P}(WWNNN) = \frac{3}{8} \cdot \frac{2}{7} \cdot \frac{5}{6} \cdot \frac{4}{5} \cdot \frac{3}{4}.$$

Outcome $WWNNN$ gives $C = 2$, but so does an outcome like $WNWNN$. In fact, there are $\binom{5}{2}$ such outcomes, so the final answer is

$$\mathbb{P}(C = 2) = \frac{3}{8} \cdot \frac{2}{7} \cdot \frac{5}{6} \cdot \frac{4}{5} \cdot \frac{3}{4} \cdot \frac{5}{2} \cdot \frac{4}{1}$$
$$= \boxed{0.5357\ldots}$$

303. THE FOREST

There are believed to be 20 deer in a forest. During one survey, 5 of the deer are tagged. During the second survey, 11 of the deer are tagged. If the tagging of the deer is random, let T be the number of deer tagged twice.
a. What is $\mathbb{E}[T]$?
b. What is $\text{sd}(T)$?

SOLUTION
a. This is $(5)(11)/20$, or $\boxed{2.750}$.
b. This is

$$\text{sqrt}\left(2.75 \cdot \frac{4 \cdot 10}{20 \cdot 19}\right),$$

or about $\boxed{0.5380}$.

MORE POISSON POINT PROCESSES

305. BACK TO THE CELLAR

In the cellar from the story the space is a polygon with vertices $(0,0), (0,12), (12,12), (12,2), (7,2)$ and $(7,0)$.
a. What is the chance that there are at least 12 rats in the cellar?
b. Given a point that marks a rat's location, what is the chance that the point has second coordinate at least 2?
c. What is the chance that there are no points with second coordinate less than 2?

SOLUTION
a. The area of the cellar is $70 + 50 = 120$. The rate of rats is 0.1 per square foot. Therefore, the number of rats in the cellar is Poisson with mean 12.

So the question is asking, what is the chance that a Poisson with rate 12 is at least 12. Let N be the number of rats. Then

$$\mathbb{P}(N \geq 12) = \sum_{i=12}^{\infty} \mathbb{P}(N = i).$$

To make this easier, use the complement rule.

$$\mathbb{P}(N \geq 12) = 1 - \mathbb{P}(N < 12)$$
$$= \sum_{i=0}^{11} \mathbb{P}(N = i)$$
$$= \sum_{i=0}^{11} \frac{\exp(-12) 12^i}{i!}$$
$$= \boxed{0.5384\ldots}.$$

b. This will be the area of the shaded part of the cellar, divided by the total area of the cellar.

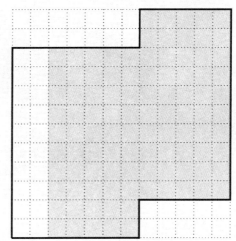

The shaded area is $5 \cdot 10 + 5 \cdot 10 = 100$ out of 120. Therefore, the chance of a point falling into the area is

$$\frac{100}{120} = \frac{5}{6} = \boxed{0.8333\ldots}.$$

c. The smaller region of area 10 forms its own Poisson point process, with mean $(10)(0.1) = 1$. So the chance of having $X = 0$ points in it is

$$\exp(-1) \frac{1^0}{0!} = \boxed{0.3678\ldots}.$$

307. THE RESTAURANT

Customers arrive to a restaurant at times modeled by a Poisson point process of rate 15 per hour. If 10 customers arrive in the first hour, what is the chance that exactly 5 customers arrive in the first half-hour?

SOLUTION

For the 10 customers that arrive, each has a 1/2 chance of arriving in the first half hour, because

$$\frac{\text{Leb}([0, 1/2])}{\text{Leb}([0, 1])} = \frac{1/2}{1} = 1/2.$$

So the number of points that fall into the smaller interval has a binomial distribution with parameters 10 (because there are 10 points total) and 1/2 (because the chance of falling in the smaller interval is 1/2.
 That is,

$$[N_{[0,1/2]} \mid N_{[0,1]}] \sim \text{Bin}(10, 1/2).$$

In particular, the chance that $(N_{[0,1/2]} = 5 \mid N_{[0,1]})$ will be

$$p = \binom{10}{5}\left(\frac{1}{2}\right)^5\left(1 - \frac{1}{2}\right)^5$$
$$= 252/2^{10} = \boxed{0.2460\ldots}.$$

309. AN ABSTRACT SPACE

Suppose S is a space with measure $\mu(S) = 15.2$. A set $A \subseteq S$ has $\mu(A) = 11.4$. Given a point $x \in S$ in the Poisson point process of constant rate $\lambda = 2.1$ over S, what is the chance that $x \in A$?

SOLUTION

This is a highly abstract problem, fortunately it is not necessary to know much about the space S to answer the question. For instance, the dimension of S is completely unknown, but that does not affect the solution!

Because A has measure 11.4, the chance that a point uniform over S lands in A is just

$$\frac{11.4}{15.2} = \boxed{0.7500}.$$

TRANSFORMING MULTIVARIATE RANDOM VARIABLES

311. ONE DIMENSIONAL TRANSFORM

Suppose X has pdf (for $x > 0$)

$$f_X(x) = \frac{1}{x\sqrt{\tau}}\exp(-\log(x)^2/2).$$

Let $Y = \log(X)$.
a. What is the density of Y?
b. What is the distribution of Y?

SOLUTION

Note $[\log(x)]' = 1/x$. Also, if $y = \log(x)$, then $x = \exp(y)$. Hence

$$f_Y(y) = 1/(1/|\exp(y)|)f_X(\exp(y))$$
$$= \exp(y)\frac{1}{\exp(y)\sqrt{\tau}}\exp(-y^2/2)$$
$$= \frac{1}{\sqrt{\tau}}\exp(-y^2/2).$$

a. The density of Y is

$$\boxed{\frac{1}{\sqrt{\tau}}\exp(-y^2/2).}$$

b. This is the density of a standard normal. So $\boxed{Y \sim \mathsf{N}(0, 1)}$.

313. TWO DIMENSIONAL TRANSFORM

Suppose $f(x, y) = x^2 y$. If $X \sim \text{Unif}([0, 1])$ and $Y \sim \text{Unif}([0, 1])$, what is the density of $S = f(x^2, y)$?

SOLUTION

In this case f takes two inputs and returns one output. For our transform method, if two inputs are taken two outputs should be produced. Fortunately, this is easy to fix. Simply create another output to go along with the first! As long as our transform is nontrivial, this will work. For instance, one could use:

$$f_1(x,y) = x^2 y$$
$$f_2(x,y) = x$$

This could be combined into one function F that says:

$$F(x,y) = (x^2 y, x).$$

Making $(S, X) = F(X, Y)$, and

$$J(x,y) = \begin{pmatrix} 2xy & x^2 \\ 1 & 0 \end{pmatrix},$$

and so the determinant is $-x^2$ and $|J| = x^2$ over the domain of $[0,1]$.

Then $F^{-1}(s, x) = (x, s/x^2)$.

Therefore, the joint density is

$$f_{(S,X)}(s,x) = \frac{1}{x^2} \mathbb{I}(x \in [0,1], y \in [0,1])$$
$$= \frac{1}{x^2} \mathbb{I}(x \in [0,1], s \in [0, x^2]).$$

To find the marginal density of x, integrate out x.

$$f_S(s) = \int_x \frac{1}{x^2} \mathbb{I}(x \in [0,1]) \mathbb{I}(s \in [0, x^2]) \, dx$$
$$= \mathbb{I}(s \in [0,1]) \int_{x=\sqrt{s}}^1 \frac{1}{x^2} \, dx$$
$$= \mathbb{I}(s \in [0,1])(-1/x)\big|_{\sqrt{s}}^1$$
$$= [s^{-1/2} - 1] \mathbb{I}(s \in [0,1]).$$

That is to say, the density is

$$\boxed{f_S(s) = [s^{-1/2} - 1] \mathbb{I}(s \in (0,1]).}$$

INDEX

ith moment, 107
σ-algebra, 13
σ-algebra of sets, 36

average, 68

Bernoulli distribution with parameter p, 45
Bernoulli point process with parameter p, 48
Bernoulli process with parameter p, 48
beta distribution with parameters a and b, 116
beta function, 117
binomial coefficients, 34
binomial random variable with parameters n and p, 49

cdf, 43
centered, 90
collection, 3
complement, 7
conditional probability formula, 20
contains, 3
continuous, 37
convolution, 102
correlation, 96
countable logical AND, 7
countable logical OR, 7
covariance between X and Y, 91
covariance matrix, 113
cumulative distribution function, 43

density, 59, 64, 84
discrete, 34
disjoint, 2
distribution, 33

element, 3
event, 3
expectation, 68
expected value, 68, 69
exponential distribution with rate λ, 45

first moment, 107

gamma distribution with parameters r and λ, 55

generating function, 104
geometric random variable with parameter p, 50

half integer, 109

iid, 36
implies, 10
independent, 24–26, 36
independent and identically distributed, 36
indicator function, 1
indicator random variable, 45
infimum, 48
inner product norm, 92
integrable, 69
intersection, 39

Jacobin matrix of F, 143

linear operator, 70
logical AND, 7
logical NOT, 7
logical OR, 6
logical statement, 1

marginal density, 86
mean, 68
measurable with respect to X, 33
median, 65
median set, 65
mode set, 64
moment, 107
moment generating function, 107
multinormal, 113
multivariate normal random variable, 113
mutually exclusive, 2

negation, 7
negative binomial random variable with parameters r and p, 50
norm, 91
normal random variable with mean μ and variance σ^2, 111
normalized density., 60
normalizing constant, 60

odds, 22

partition, 15
pdf, 59, 64
pmf, 64
Poisson distribution with parameter μ, 55
Poisson point process, 139
Poisson point process on $[0, \infty)$ of rate λ, 54
polar coordinates, 144
posterior, 31
Principle of Indifference, 3
prior, 31
probability density function, 59
probability distribution, 13
probability function, 13
probability mass function, 64
probability measure, 13

Radon-Nikodym derivative, 59
random variable, 33
real valued inner product, 91
real-valued random variable, 46
relative error, 126
relative standard deviation)' also known as 'r term(coefficient of variation, 127
relative variance, 127

set, 3
shift equivalent, 93
standard deviation, 92
standard exponential distribution, 45, 54
standard normal distribution, 108
standard uniform, 41
stochastic process, 48
subset, 10
survival function, 67

uncorrelated, 97
uniform measure over $[a, b]$, 41
uniform measure over A, 39
unnormalized density, 60

variance, 92
vector space, 90

with respect to Lebesgue measure, 59

Made in the USA
Las Vegas, NV
18 February 2025

3602c624-8bfe-41e8-976c-942728e0ffd9R01